Current Topics in Microbiology
209 and Immunology

Springer-Verlag Berlin Heidelberg GmbH

Bacterial Invasiveness

Edited by V.L. Miller

With 16 Figures

Springer

Professor Virginia L. Miller, Ph.D.
Department of Microbiology and Molecular Genetics
University of California, Los Angeles
405 Hilgard Avenue,
Los Angeles CA 90095, USA

Cover illustration: The Legionnaires' disease bacterium, Legionella pneumophila, has an intracellular fate that is determined by the icmA locus. Upon infection, wild-type L. pneumophila Philadelphia-1 is placed in a unique phagosome that does not fuse with host lysosomes. Therefore, phagosomes containing fluoresceinated wild-type L. pneumophila (green) show no colocalization with host lysosomes prelabelled with rhodamine-dextran (red) in macrophage-like U937 cells (Figure 1). In contrast, L.pneumophila mutant 25D contains an icmA mutation and can not prevent phagosome-lysosome fusion. Digestion of fluoresceinated mutant bacteria within phagolysosomes and dispersion of its contents generate many punctated yellow vesicles containing the colocalized bacterial and lysosomal markers (Figure 2). For both images, macrophage-like U937 cells were incubated for 36 hours with the lysosomal marker lysine-fixable rhodamine-dextran (70,000 MW), washed and then infected with L. pneumophila that had been previously labelled by a succinimidyl ester of fluorescein. After incubating 6 hours for the wild-type and 16 hours for the mutant L-pneumophila, the cells were fixed and imaged by fluorescent confocal microscopy. Confocal fluorescent images by Lawrence A. Wiater.

Cover design: Künkel+Lopka, Ilvesheim

ISBN 978-3-642-85218-3   ISBN 978-3-642-85216-9 (eBook)
DOI 10.1007/978-3-642-85216-9

© Springer-Verlag Berlin Heidelberg 1996
Library of Congress Catalog Card Number 15-12910
Softcover reprint of the hardcover 1st edition 1996

The use of general descriptive names, registered names, trademarks, etc. in this publication does not imply, even in the absence of a specific statement, that such names are exempt from the relevant protective laws and regulations and therefore free for general use.

Product liability: The publishers cannot guarantee the accuracy of any information about dosage and application contained in this book. In every individual case the user must check such information by consulting other relevant literature.

Typesetting: Thomson Press (India) Ltd., Madras
SPIN: 10495427   27/3020/SPS – 5 4 3 2 1 0 – Printed on acid-free paper

Preface

Intracellular pathogens are responsible for a number of important diseases world wide, including such devastating ones as malaria and tuberculosis. Many of these pathogens have been difficult to study because either they are obligate intracellular pathogens or the genetic tools to dissect the process have not been available. This volume focuses on those intracellular pathogens that have been studied most extensively at the molecular, genetic, and cellular levels. The reviews attempt to integrate the information derived from these diverse approaches into a cohesive picture. While there have been a number of reviews on various aspects of this topic, it is useful to look at them together, as this highlights the similarities and differences. This series includes a similar compilation of reviews published in 1988; looking back at that issue, it is striking to see how rapid the progress of research in this area has been. At that time, the general steps of entry had been described and a few of the genes involved had just been identified. Now we find that a description of the entry steps is being laid out at the molecular level and the important signal transduction events are being elucidated. In addition, numerous genes encoding products involved in this process have been identified and sequenced, and the mode of action of their products is being intensively investigated.

When first studied, the entry steps taken by these diverse bacteria seemed to be quite similar, while the genetic basis of entry seemed to be dissimilar. As we have learned more about each aspect, though, we find both similarities and differences at all levels. *Shigella*, *Yersinia*, and *Salmonella* probably all target M cells as a site of entry. However, *Yersinia* probably enters due to the interaction of invasin on the bacterial cell surface with β_1 integrin receptors on the eukaryotic cell. Uptake of *Legionella pneumophila* also occurs via an integrin receptor (CR1 and CR3). Enteropathogenic *Escherichia coli* (EPEC) synthesizes intimin, a protein related to invasin, which mediates adherence and probably plays a role in entry as well. Although, not related to invasin at the sequence level, internalin produced by *Listeria monocytogenes* may act in an analogous manner, but further studies are required before this conclusion can be made. No ligand comparable to invasin has been identified for *Shigella* and *Salmonella*, which appear instead to enter by stimulating macropinocytosis. While both *Shigella* and *Salmonella* stimulate membrane ruffling by triggering signal transduction events involving host protein tyrosine phosphorylation, the host pathways involved differ. Despite this, recent data suggest that the bacterial signaling proteins synthesized by *Shigella* and *Salmonella* may be closely related. Curiously, type-III secretion pathways have been implicated as important for the virulence of *Yersinia*, *Salmonella*, *Shigella*, and EPEC. Nevertheless, the proteins secreted by these pathways do not play the same role for each species. In the case of *Shigella*, *Salmonella*, and EPEC the secreted proteins are probably involved in triggering the signal transduction cascade in the eukaryotic cell, leading to entry. In contrast, the secreted proteins of *Yersinia* are involved in blocking host defense activity and may be antiphagocytic.

Once inside the cell, the subsequent steps observed for the gram-negative pathogen *Shigella flexneri* and the gram-positive pathogen *L. monocytogenes* are remarkably similar. They both escape the phagocytic vacuole, multiply freely in the cytoplasm, and induce polymerization of host actin—a process that results in direct cell-to-cell spread. Although many of the bacterial products involved have been identified, no similarities at the sequence level have been noted. However, further research into the structure and mode of action of these proteins may reveal evolutionary relationships not evident from the primary sequence.

Many of the results and ideas presented in these reviews were only recently published or have yet to be published, and thus represent the very latest in a rapidly moving field. Although much has been learned about bacterial invasion in the past decade, we still do not have a complete story. In fact, the more we learn, the more the imagination is teased and the more elegant the solutions to tough problems encountered by the bacteria seem. Clearly, many interesting discoveries remain to be made.

Los Angeles, USA   VIRGINIA L. MILLER

List of Contents

List of Contributors

(Their addresses can be found at the beginning of their respective chapters.)

COSSART, P. 61
DONNENBERG, M.S. 79
DRAMSI, S. 61
GALÁN, J.E. 43
HORWITZ, M.A. 99

ISBERG, R.R. 1
LEBRUN, M. 61
PARSOT, C. 25
SANSONETTI, P.J. 25
SHUMAN, H.A. 99

Uptake of Enteropathogenic *Yersinia* by Mammalian Cells

R.R. Isberg

1 Introduction

The enteropathogenic *Yersinia* are two gram-negative bacterial species related to the causative agent of bubonic plague, *Yersinia pestis*. *Yersinia enterocolitica* causes a variety of intestinal diseases as well as mesenteric lymphadenitis, with outbreaks occurring primarily in cold-weather climates (Anderson et al. 1991; Tripoli et al. 1990). The most notable characteristic of *Y. enterocolitica* infections is that mild or inapparent bacterial infections may trigger a number of autoimmune disorders (Saario et al. 1992; Toivanen et al. 1993). These include thyroiditis and reactive arthritis, particularly in individuals harboring the HLA B27 histocompatibility allele (Maki et al. 1991; Toivanen and Toivanen 1994; Tomer and Davies 1993; Wenzel et al. 1991). *Y. pseudotuberculosis* causes enteric diseases and associated complications very similar to those seen with *Y. enterocolitica* (Fukushima 1991; Fukushima et al. 1990; Stahlberg et al. 1987).

Howard Hughes Medical Institute and Department of Molecular Biology and Microbiology, Tufts University School of Medicine, 136 Harrison Ave., Boston, MA 02111, USA

Enteropathogenic *Yersinia* diseases have been studied extensively to gain insight into several important pathogenic processes. First, they have been analyzed to understand how a pathogen can spread from an intestinal site to set up a replicative niche in other tissue sites (CARTER 1975; HEESEMANN et al. 1993). Second, they express a number of proteins that appear to be analogs of well-characterized regulatory factors present in the mammalian hosts (STRALEY et al. 1993). Analysis of these proteins is certain to provide insight into how disruption of regulatory processes within host cells contributes to establishing and maintaining an infectious disease. Third, as they are strongly associated with the presence of reactive arthritis is susceptible individuals, their study is hoped to uncover basic events within the host that lead to rheumatological disorders (MAKI et al. 1992). Finally, they have allowed dissection of bacterial uptake by normally nonphagocytic cells (ISBERG and TRAN VAN NHIEU 1994a). The focus of this chapter will be on this last aspect of *Yersinia* pathogenesis, with some overview of the pathogenesis of these microorganisms.

2 Enteropathogenic *Yersinia* Disease

Disease caused by enteropathogenic *Yersinia is* is initiated by ingestion of contaminated foodstuffs, generally due to undercooked meats or contaminated lots of milk or water supplies (FUKAI and MARUYAMA 1979; MERILAHTI et al. 1991). The most extensively studied animal models for this enteric infection are the rabbit and mouse, in which the disease proceeds by translocation across the epithelium of the ileum or the colon to the submucosal region (CARTER 1975). In the mouse ileum, there is some evidence that *Y. enterocolitica* is found within a subset of cells called M cells, located over the lymphoid Peyer's patches (GRUTZKAU et al. 1990; HANSKI et al. 1989a). In the case of *Y. pseudotuberculosis*, uptake into cells overlaying the Peyer's patches occurs rapidly after injection of bacteria into the lumen of ligated intestinal loops, internalized bacteria being evident within 15 min after their introduction into the loop (MARRA and ISBERG unpublished). Internalization appears rather patchy, with some intestinal cells showing large numbers of bacteria and adjoining cells showing none, indicating that the mammalian cell receptors for the bacteria may be unevenly distributed through the intestine. Bacteria can be found within the Peyer's patch shortly after this time, with microorganisms both within and outside cells. In animal models, enteropathogenic *Yersinia* drain into mesenteric lymph nodes, and large numbers of bacteria can be found in these glands within 24 h of initial infection (CARTER 1975). In human infections, it appears that the disease terminates at infection of mesenteric lymph nodes (O'LOUGHLIN et al. 1990). After oral infection, susceptible animals, such as many mouse strains, are unable to harness the replication of enteropathogenic *Yersinia* at this site, and bacterial growth continues after drainage into the liver and spleen. The bacteria are exclusively extracellularly localized and continue to be outside host cells for the duration of the disease. Replication at these sites eventually causes death of the animal.

The most striking characteristic of the disease is that during fulminant replication within host tissues, the microorganism appears to actively resist uptake by both professional phagocytes and normally nonphagocytic cells (SIMONET et al. 1992). This situation is quite different from what is seen in the early stages of the infection and from what is seen after bacteria grown at ambient temperature are introduced onto mammalian cells (ISBERG 1989). This extracellular localization is probably a direct result of factors produced by a large virulence plasmid found in all *Yersinia* species (STARLEY et al. 1993). Encoded on the plasmid are three sets of factors involved in proper deposition of virulence-associated proteins, either within host cells or into the extracellular milieu. One group of factors, the *lcr* or *vir* loci, is involved in regulation of the synthesis and proper localization of extracellular virulence proteins. The products of the *lcr* lock are responsible for ensuring that virulence-associated proteins expressed from the plasmid are maximally expressed at physiological temperatures under Ca^{2+}-limiting conditions (HOE et al. 1992; RIMPILAINEN et al. 1992). Their expression also results in a phenomenon known as Ca^{2+}-dependent growth, causing bacteria to die when incubated at 37°C in bacteriological media lacking Ca^{2+} (FOWLER and BRUBAKER 1994). A second group of factors probably provides the machinery for export of virulence-associated proteins that lack N-terminal signal sequences (FIELDS et al. 1994; HADDIX and STRALEY 1992; MICHIELS et al. 1991). Many of these loci corresponding to *vsc* and *svc* genes, as well as some of the *lcr* genes, show strong sequence similarity to the loci found in numerous bacterial pathogens of plants and animals that express proteins involved in export and assembly of a variety of virulence-associated secreted proteins (ALLAOUI et al. 1993; HUANG et al. 1992). The targets of this specialized secretion machinery are the Yops, which constitute the third important set of proteins expressed by the virulence plasmid (CORNELIS et al. 1989; STRALEY and BOWMER 1986). Although the functions of many of the Yops are unknown, several have been identified either as acting directly on or within host cells, or as being involved in the deposition of *Yersinia* proteins within target cells (BLISKA et al. 1991; GALYOV et al. 1993).

The extracelluar locale of enteropathogenic *Yersinia* replicating within the liver and spleen is probably a direct result of multiple phenomena. It is most probably the result of deposition of two Yop proteins within the host cell cytosol that antagonize bacterial uptake (ROSQVIST et al. 1991). One of these, the product of the *vopE* gene, is referred to as a cytotoxin, with the function of inducing depolymerization of the host cell cytoskeleton (ROSQVIST et al. 1989, 1991). As most phagocytic processes require cytoskeletal rearrangements, this is probably the cause of the disruption of bacterial uptake. The second antiphagocytic protein is the product of the *vopH* gene, which is tyrosine phosphatase (ROSQVIST et al. 1988a; 1990). It prevents the formation of a phosphorylated protein intermediate that is an important signal for phagocytosis (BLISKA et al. 1991). The role of this protein in interfering with uptake will be discussed in greater detail below.

3 Uptake of Enteropathogenic *Yersinia* by Cultured Cell Lines and the Identification of Invasin

Morphological studies of animal infection models indicate that enteropathogenic *Yersinia* are found in an intracellular niche during the period directly following the initial bacterial encounter with the intestinal mucosa (GRUTZKAU et al. 1990; HANSKI et al. (1989b). Intracellular localization during an infection probably allows translocation into deeper tissues. Most adherent cultured cells lines efficiently internalize enteropathogenic *Yersinia* (DEVENISH and SCHIEMANN 1981). Uptake into cultured cells occurs via circumferential binding of the host cell about the surface of the bacterium, with individual microorganisms internalized into membrane-bound phagosomes (BOVALLIUS and NILSSON 1975). Maximal uptake is observed if the bacteria are cultured at temperatures lower than 28°C prior to incubation with the mammalian cells (ISBERG et al. 1988), and uptake levels that approach 50% of the initial inoculum are possible if bacteria are cultured optimally. Preculturing at 37°C, on the other hand, results in reduced uptake efficiency due to induction of the antiphagocytic YopE and YopH proteins (Rosqvist et al. 1988a, 1990). The fact that uptake occurs maximally in the absence of the *Yersinia* virulence plasmid indicates that the primary factors associated with efficient uptake are chromosomally encoded (ISBERG and FALKOW 1985).

Identification of *Yersinia* factors required for uptake was accomplished by selecting for cosmids that contain *Y. pseudotuberculosis* chromosomal DNA able to confer the ability of innocuous *Escherichia coli* laboratory strains to enter cultured mammalian cells (ISBERG and FALKOW 1985). Such molecular clones, all of which expressed the product of the *inv* gene, invasin, were selected based on their ability to allow survival of the bacteria in mammalian cell cultures treated with gentamicin, an antibiotic unable to efficiently kill intracellular bacteria (MANDELL 1973). The product of *inv*, predicted to be 986 amino acids, was shown to be exposed on the bacterial cell surface (ISBERG et al. 1987). Bacteria that express invasin bind tightly to the mammalian cell surface (ISBERG et al. 1987). Mammalian cells adhere to invasin that has been fractionated on an SDS-polyacrylamide gel and transferred to Immobilon filters, or to purified invasin derivatives immobilized on plastic dishes (ISBERG and LEONG 1988). Adhesion requires divalent cations and is localized in the carboxyl terminal of the protein, as purified hybrid proteins containing the carboxyl terminal 192 amino acids of invasin fused to *E. coli* maltose-binding protein support mammalian cell adhesion (LEONG et al. 1990). An amino terminal region extending over 600 amino acids is required for proper localization and presentaiton of the cell adhesion domain of invasin on the bacterial cell surface.

The *inv* gene is highly similar to several loci associated with bacteria-host cell interactions that have been found in a variety of related gram-negative enteric pathogens, such as enteropathogenic *E. coli* and *Citrobacter freundii* (Table 1; JERSE et al. 1990; SCHAUER and FALKOW 1993; YU and KAPER 1992). Some of these other loci express products that have been termed intimins, based on the fact that

Table 1. Gene families associated with Yersinia uptake into cultured cells

Gene	Microorganisms	Function	Reference
A. Ail family			
ail	Y. Enterocolitica	Adhesion, uptake complement resistance	Miller and Falkow (1988) Bliska and Falkow (1992)
	Y. pseudotuberculosis	Complement resistance	Yang and Isberg unpublished
pagC	S. typhimurium	Intracellular survival	Pulkkinen and Miller (1991)
rck	S. Typhimurium	Complement resistance Adhesion, uptake	Haffernan et al. (1992) Haffernan et al. (1994)
ompX	E. cloacae	?	Stoorvogel et al. (1991)
	E. coli	?	Mecsas et al. (1995)
lom	bacterophage λ	?	Munn and Reeves (1985)
B. Invasin family			
inv	Y. pseudotuberculosis	Adhesion, uptake, deposition of Yops	Isberg and Falkow (1985) Rosquist et al. (1990)
	Y. enterocolitica	Adhesion, uptake	Miller and Falkow (1988)
eaeA	Enteropathogenic E. coli	Attachment and actin pedestals	Jerse et al. (1990)
	Enterohaemorrhagic E. coli	Attachment and actin pedestals	Yu et al. (1992)
	Citrobacter freundii biotype 4280	Colonic hyperplasia pedestals	Schauer and Falkow (1993)
	Hafnia alvei	Attachment and pedestals	Franker et al. (1994)

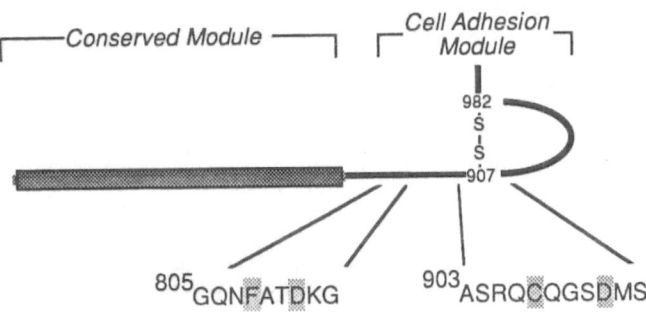

Fig. 1. Structure of the *Yersinia pseudotuberculosis* invasin protein. *Shaded region* extending approximately 700 amino acids from the amino terminal end is region with high sequence similarity to the intimins. The carboxyl terminal 192 amino acid cell-binding domain is denoted in *black*. Note the disulfide bond between amino acids 907 and 982 and the sequences of the two regions of the protein that, when mutated, exhibit defects in cell adhesion. Shaded residues, when changed to other amino acids, cause either total elimination of cell adhesion (C907 and D911) or substantial loss of adhesion (F808 and D811)

they are involved in intimate interaction of the bacterium with the host cell (DONNENBERG et al. 1993). Unlike invasin, intimins act in coordination with other bacterial factors to promote an attaching and effacing structure, in which bacteria attach to intestinal epithelial cells and cause aggregation and immobilization of host cell actin about the site of bacterial binding, with simultaneous effacement of surrounding microvilli. The region of invasin that is similar to the intimins is limited to the amino terminal 700 amino acids responsible for surface presentation of the cell adhesion domain (Fig. 1; LEONG et al. 1990). The C-terminal region of approximately 200 amino acids forms a cell adhesion module that varies in sequence among each of the homologues (YU and KAPER 1992; FRANKEL et al. 1994). Interestingly, the only striking similarity within this C-terminal region is the presence of a pair of cysteines separated by 72–78 amino acids (Fig. 1). These two cysteines form a disulfide bond in invasin that is essential for maintenance of a binding-competent conformation (LEONG and ISBERG 1993).

4 Identification and Analysis of the *ail* Gene Product

A chromosomal gene bank from *Y. enterocolitica* was introduced into an *E. coli* laboratory strain, and selection identical to that described for *Y. pseudotuberculosis* was performed to identify factors involved in uptake (MILLER and FALKOW 1988). A gene corresponding to an *inv* homologue was isolated by this procedure, as well as a second locus called *ail*, which had not been isolated in the selection from the *Y. pseudotuberculosis* gene bank (ISBERG and FALKOW 1985). *E. coli* strains harboring plasmids encoding the 178 amino acid product of the *ail* gene adhere efficiently to a variety of mammalian cell lines (MILLER et al. 1990). On the other hand, with the exception of Chinese hamster ovary cells, uptake is rather inefficient compared with that seen with the parental *Y. enterocolitica* strain (MILLER et al. 1990). The product of the *ail* locus appears to promote serum resistance when expressed in *E. coli,* and *Y. enterocolitica* strains lacking the virulence plasmid and harboring insertion mutations in *ail* locus are exquisitely sensitive to human complement (BLISKA and FALKOW 1992; PIERSON and FALKOW 1993).

The *ail* gene from *Y. pseudotuberculosis* was recently isolated using low-stringency hybridization with a *Y. enterocolitica ail* probe (Y. YANG, J. MERRIAM, and R. ISBERG, unpublished data). When expressed in *E. coli*, the *Y. pseudotuberculosis* homologue confers high-level resistance to the action of human complement but shows no apparent ability to promote bacterial adhesion or uptake. Consistent with these results is the analysis of insertion mutations in the *Y. pseudotuberculosis* chromosome that disrupts *ail* gene product function. Such mutants are indistinguishable from wild-type bacteria in their ability to adhere to cultured cells but are extremely sensitive to the bactericidal effects of serum (Y. Yang, unpublished data).

The predicted products of the *ail* and *inv* genes show no apparent homology to each other. The product of *ail*, however, is a member of a large family of genes that are found in most enteric bacteria (Table 1; MILLER et al. 1990). Some of these loci, such as *rck* in *Salmonella typhimurium*, can also contribute to complement resistance (HEFFERNAN et al. 1992). Others, such as the product of the

S. typhimurium pagC locus, which is involved in survival within phagocytes (PULKKINEN and MILLER 1991), have no identified biochemical activity. That a small reading frame can give rise to a diverse set of functions was explained by the structural model presented by Miller and co-workers based on their work with the *Y. enterocolitica ail* gene (BEER and MILLER 1992). They isolated a series of *ail* genes from several *Y. enterocolitica* serotypes. Several of these loci, when harbored in *E. coli*, had lowered uptake proficiency. Sequencing revealed that most of the divergence between different isolates was found in amino acids predicted to be located in loops between individual β strands. The authors proposed that the β strands corresponded to regions that span the outer membrane and that most of the sequence divergence could be predicted to be found on the surface-exposed face of the outer membrane. Consistent with this prediction is the fact that the membrane-spanning regions of most outer membrane proteins are β strands (KREUSCH and SCHULZ 1994).

The model for Ail protein, in which multiple membrane-spanning segments are embedded in the outer membrane and the biological activity is limited to surface-exposed loops, indicates that structures protruding from the outer surface of the bacterium could interfere with recognition of Ail protein by mammalian cells. Recent work from Pierson is consistent with this interpretation, as conditions that favor the assembly of O-antigen on the bacterial cell surface result in lowered ability of Ail to promote internalization within mammalian cells (PIERSON 1994). Furthermore, mutations that result in enhanced uptake of *Y. enterocolitica inv* mutants have disruptions in O-antigen synthesis. Uptake of the revertants requires the presence of an intact *ail* gene (PIERSON 1994).

The fact that the *ail* product has multiple activities means that its biologically relevant function is unclear. Epidemiological evidence suggests that the protein is important for pathogenesis, as there is a very close correlation between the presence of the *ail* gene and the potential of a *Y. enterocolitica* isolate to cause disease (MILLER et al. 1989). Environmental isolates that have no demonstrated potential to cause disease consistently fail to harbor the *ail* gene, whereas all known *Y. enterocolitica* isolates from human disease cases express the gene product (MILLER et al. 1989). Unfortunately, *Y. enterocolitica ail* mutants have rather subtle phenotypes. For instance, such mutations result in only a small decrease in uptake by cultured mammalian cells relative to wild type, as might be expected for microorganisms harboring an intact *inv* gene (PIERSON and FALKOW 1993). They also have little effect in animal infection models in which death is used as an end point (M. WACHTEL and V. MILLER, personal communication). This latter result may be a reflection of using animals that are exquisitely sensitive to *Yersinia* enteropathogenesis. Mouse infections may not give insight into the rather attenuated diseases that occur in human beings, especially given the close correlation between presence of the *ail* gene and pathogenic potential of the microorganism.

5 Alternate Strategies that Allow Uptake and Adhesion to Cultured Mammalian Cells

The analysis of Y. pseudotuberculosis inv mutants indicates that there are other factors besides invasin and Ail that allow interaction with mammalian cells. Y. pseudotuberculosis inv mutants grown at 28°C enter mammalian cells at 1% the efficiency of wild-type strains, as measured by survival after gentamicin treatment (ISBERG 1989). Although low, this level is higher than that seen for laboratory E. coli strains and higher than that for Y. pseudotuberculosis inv mutants cured of the virulence plasmid. To identify the plasmid-borne factor that promotes this low-level uptake, a gene bank was made from the virulence plasmid and introduced into a Y. pseudotuberculosis inv mutant, and strains able to enter mammalian cells were selected by introducing the entire pool onto a HEp-2 cell monolayer (YANG and ISBERG 1993). All such strains that survived contained plasmids encoding the previously characterized YadA gene. In a complementary study, it was shown that E. coli strains harboring plasmids in which YadA was placed under the control of the lac promoter were able to enter mammalian cells if the bacteria were cultured under conditions that induced expression of vadA (BLISKA et al. 1993). Defined Y. pseudotuberculosis vadA inv double mutants were shown to be as defective for uptake as laboratory E. coli strains, indicating that these two were the primary factors responsible for uptake into the cultured lines analyzed (YANG and ISBERG 1993).

The product of vadA is a 435 amino acid primary translation product that makes fibrillar structures on the surface of the bacterium (SKURNIK and WOLF-WATZ 1989). It had been previously characterized as an adhesive molecule with a variety of activities, including binding to mammalian cells. In addition, it binds extracellular matrix proteins such as collagen (EMODY et al. 1989) and fibronectin produced by cultured cells [but not that derived from plasma; (SCHULZE-KOOPS et al. 1993)] and components of mucus, and it provides complement resistance and antagonism of phagocytosis by neutrophils (CHINA et al. 1993, 1994). YadA-promoted adhesion to mammalian cells may involve binding to cellular fibronectin or collagen or direct binding to a cellular receptor. YadA-dependent adhesion is inhibited by antibodies directed against β_1 integrin receptors (BLISKA et al. 1993). This indicates either that YadA binds directly to these receptors, or that mammalian proteins able to bind these receptors form a bridge between the bacterium and the mammalian cell. As fibronectin and collagen bind both YadA and integrin receptors, the formation of a bridge by this extracellular matrix (ECM) seems the most likely model for the adhesion event that precedes YadA-dependent uptake.

The biological relevance of uptake promoted by YadA is rather unclear. The protein obviously has a plethora of activities, most of which have little relationship to uptake. Furthermore, the efficiency of uptake promoted by YadA is unimpressive relative to that seen with the other characterized entry pathways. It seems most likely that uptake is a consequence of efficient binding of the bacteria to ECM proteins, which in vivo are immobilized in a rather inflexible matrix.

In cultured cell lines, however, fragments of matrix bound to the mammalian cell may bind the bacteria, and the resulting adhesion complex may be internalized at a low level.

Analysis of *vadA inv* mutants showed that there is yet another activity expressed by *Y. pseudotuberculosis* that allows adhesion to mammalian cells. *Y. pseudotuberculosis inv* mutants cured of the virulence plasmid adhere efficiently to mammalian cells, dependent on the temperature at which the bacteria are cultured prior to incubation with mammalian cells. Bacteria pregrown at 28°C adhere much less efficiently than those grown at 37°C (ISBERG 1989), and BLISKA noted that culturing at pH =6.0 prior to binding of mammalian cells vastly improves the adhesion (J. BLISKA, personal communication). As these were conditions in which a pilus-like structure called pH 6 antigen (psaEAB locus) is maximally expressed (LINDLER and TALL 1993), he hypothesized that this factor was responsible for the fourth adhesion pathway (J. BLISKA, personal communication). Analyses of *Y. pseudotuberculosis inv vadA psaA* triple or *inv YadA psaA ail* quadruple mutants indicate that elimination of *inv YadA* and *psaA* is sufficient to prevent detectable adhesion to cultured HEp-2 cells (Y. YANG and R. ISBERG unpublished).

One important aspect of adhesion via the pH 6 antigen structure is that, although it promotes efficient extracellular binding of the bacteria to the mammalian cells, no detectable uptake occurs. This illustrates a central feature of bacterial adhesion to host cells: mere binding of a bacterium to a host cell is not sufficient to result in internalization of the microorganism. There must be determinants other than simple binding that are necessary to facilitate uptake, or else the adhesion event must meet certain requirements before internalization can occur. Analysis of the mode of action of invasin has helped to clarify this particular problem.

6 Mechanism of Action
of the *Y. pseudotuberculosis* Invasin Protein

As illustrated in the pH 6 antigen example, the ability to bind mammalian cells is not sufficient to promote entry. To determine whether binding to a special receptor on the mammalian cell surface is the key element in allowing invasin-promoted uptake, proteins that bind invasin were identified (ISBERG and LEONG 1990). Detergent extracts of surface-labeled mammalian cells from a variety of cell lines or tissue sources were subjected to affinity chromatography on invasin-agarose columns, and bound proteins were eluted with EDTA (ISBERG and LEONG 1990). Five different proteins were isolated by this procedure that were members of the integrin superfamily of cell adhesion molecules (Table 2; ISBERG and LEONG 1990). The integrin family consists of more than 20 αβ heterodimeric surface proteins involved in binding to ECM proteins and in cell-cell interactions. There are

Table 2. Invasin receptors and their host ligands

Integrin	Typical cells	Typical ligands
$\alpha_v\beta_1$	Neuroblastoms	Vitronectin, fibronectin
$\alpha_3\beta_1$	Epithelial Cells	Fibronectin, epiligrin
$\alpha_4\beta_1$	Lymphocytes, monocytes	Fibronectin, VCAM-1
$\alpha_5\beta_1$	Endothelial cells, Fibroblasts	Fibronectin
$\alpha_6\beta_1$	Platelets	Laminin

over a dozen α chains and eight β chains (β_1–β_8) that can assort with each other to form at least 20 different heterodimers (HYNES 1992). The substrates bound by each receptor are determined by the particular chains present in the heterodimer. For instance, the integrin $\alpha_5\beta_1$ is a receptor for fibronectin (ARGRAVES et al. 1986), whereas $\alpha_v\beta_3$ is a vitronectin receptor (LAM et al. 1990).

Each of the integrin receptors bound by invasin has the β_1 chain (Fig. 2; ISBERG and LEONG 1990), but invasin does not bind directly to this chain, as determinants on both the α and β_1 chains are required for binding. The isolated β_1 chain cannot bind invasin (G. TRAN VAN NHIEU, unpublished), nor can the purified collagen receptor $\alpha_2\beta_1$ (ISBERG and LEONG 1990). The identified integrins are true receptors for invasin, as their intact function is required for uptake. For instance, function-

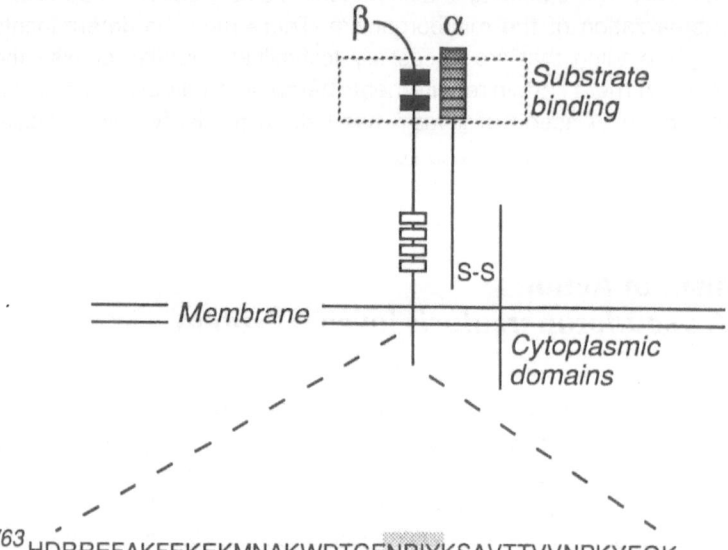

^{763}HDRREFAKFEKEKMNAKWDTGENPIYKSAVTTVVNPKYEGK

Fig. 2. Structure of integrin invasin receptors. Shown is typical $\alpha\beta$ heterodimer that binds invasin. Several invasin receptors have a cleaved C-terminal α chain disulfide-link to the ligand binding region of the α chain. Shown are the seven amino terminal repeats (■) containing three or more divalent cation binding sites corresponding to EF hand-like sequences in the α chain. The cysteine-rich regions in the β chain are noted (▭), as are ther regions involved in ligand recognition (■). The cytoplasmic domain of the β chain is displayed showing the critical shaded NPIY sequence

blocking antibodies directed against the β_1 chain prevent both invasin-dependent bacterial binding to the mammalian cell and uptake (Isberg and Leong 1990). Uptake can also be prevented by a combination of anti-α monoclonal antibodies (mAbs) that recognize the appropriate receptors on a particular cell type. Finally, a cell line lacking the β_1 chain is unable to bind or internalize bacteria via the invasin pathway (E. Krukonis and R. Isberg, unpublished data).

Integrin receptors are able to couple extracellular adhesion to numerous signaling events within the mammalian cell (Hynes 1992), some of which allow association of the β_1 chain to cytoskeletal-associated proteins such as talin (Horwitz et al. 1986), α-actinin (Otey et al. 1990), and the focal adhesion kinase p125^{FAK} (Schaller et al. 1992). Most notable among the responses to integrin-mediated adhesion are a variety of phosphorylation events (Guan et al. 1991), rearrangement of the cytoskeleton (Hynes 1992), and alteration of cytoplasmic pH (Schwartz et al. 1991). In addition, it has been known for a number of years that adhesion of integrins can be modulated by intracellular signaling events such as those induced by small molecule mediators or by cellular adhesion to appropriate substrates (Ennis et al. 1993; Hermanowski-Vosatka et al. 1992).

The association of cytoskeletal elements with the β integrin chain may be important for invasin-promoted uptake (Finlay and Falkow 1988). Internalization appears to require rearrangement of the cytoskeleton and is inhibited by agents that depolymerize actin, such as cytochalasin D (Finlay and Falkow 1988). The presence of cages of actin surrounding phagosomes as determined by fluorescence microscopy has been reported, suggesting direct attachment of cytoskeletal elements to some factor located in the phagosomal membrane (Young et al. 1992). One interpretation of these results is that attachment of the bacterium to the integrin receptor allows an important functional interaction to occur that signals rearrangement of the cytoskeleton. If this is the case, then it is possible that mere attachment of the bacterium to the integrin receptor is sufficient to promote uptake.

Several lines of evidence indicate that integrin ligands such as fibronectin promote rather inefficient uptake of bound bacteria, so simple attachment of a bacterium to an integrin is not sufficient to result in bacterial uptake. A variety of bacteria have fibronectin-binding proteins on their surface (Flock et al. 1987; Hook et al. 1989), yet fibronectin-coated bacteria bound to the mammalian cell are internalized poorly (as is the case with YadA-promoted uptake; Yang and Isberg 1993). Furthermore, *Staphylococcus aureus* bacteria coated with identical quantities of invasin and fibronectin have vastly different fates, the invasin-coated bacteria being much more efficiently internalized (Rankin et al. 1992; Tran Van Nhieu and Isberg 1993a).

In spite of these results, invasin does not have a function distinct from adhesion that triggers internalization of the bound bacterium. This was demonstrated by analyzing bacterial uptake promoted by successively smaller derivatives of MBP-invasin hybrid proteins. Bacteria coated with the smallest derivatives capable of binding mammalian cells were efficiently internalized, whereas binding-defective derivatives promoted no internalization (Rankin et al. 1992). Although this result does not rule out the possibility that invasin has overlapping binding and uptake activities in this domain, the simplest interpretation is that integrin binding is the only activity present in invasin that is necessary for uptake.

One explanation for the different responses to invasin and fibronectin is that ligands that promote extracellular localization of bacteria bind a different site on the integrin receptor than does invasin. This possibility was eliminated by analyzing the interaction of invasin with the $\alpha_5\beta_1$ integrin purified from human placenta (TRAN VAN NHIEU and ISBERG 1991). Fibronectin competitively inhibited the binding of labeled invasin to filter-immobilized $\alpha_5\beta_1$ integrin in this study (TRAN VAN NHIEU and ISBERG 1991). A peptide from fibronectin (GRGDSP) that inhibits binding to the $\alpha_5\beta_1$ integrin also inhibited the binding of labeled invasin to the purified receptor. Furthermore, all mAbs that block invasin binding to its receptor also block fibronectin binding. Consistent with these results were those of a related study, in which a point mutation in the β_1 chain was shown to simultaneously eliminate binding to invasin and all other characterized substrates (TAKADA et al. 1992). There must be some physical property of the binding step, however, that distinguishes invasin from other integrin ligands.

Invasin binds its integrin receptors with an extremely high affinity that is one or two orders of magnitude greater than that for most mammalian integrin substrates (TRAN VAN NHIEU and ISBERG 1991). The only other known integrin ligands that bind so avidly belong to a family of small snake venom proteins, called disintegrins, that target the platelet integrin ligand $\alpha_{IIb}\beta_3$ (NIEWIAROWSKI et al. 1994). Invasin binds to the $\alpha_5\beta_1$ integrin with a $K_d = 5 \times 10^{-9}$ M and a rather slow off rate of $t_{1/2} = 20$ min. In contrast, the K_d of fibronectin binding to the identical receptor is approximately 7×10^{-7} M (AKIYAMA and YAMADA 1985).

The large difference in affinity between fibronectin and invasin is probably the basis for their different efficiencies in promoting uptake. In support of this hypothesis, MAbs directed against the $\alpha_5\beta_1$ integrin were used as surrogate ligands for bacterial uptake (TRAN VAN NHIEU and ISBERG 1993b). For this purpose, S. aureus was coated with each of the MAbs and used to challenge cultured HEp-2 cells. All of the MAbs were able to promote similar levels of adhesion to the cell line, but the uptake efficiency was a direct linear logarithmic function of the relative affinity of the MAbs for the receptor. Overproduction of receptor allowed S. aureus coated with low-affinity MAbs to be internalized, although the relative uptake efficiency of each MAb was still dependent on affinity (TRAN VAN NHIEU and ISBERG 1993a).

There are several complementary explanations for the requirement for high-affinity binding by invasin. The need for high-affinity interaction indicates that competing events may interfere with uptake, as integrin receptors are able to bind ECM proteins as well as invasin. As a result, receptors are immobilized, reducing the number of receptors that can be used for uptake. This also means that the number of receptors available for uptake will be modulated by the affinity of binding to ECM. A second implication of the competition model is that host cells limited in receptor concentration limit the bacteria to extracellular adherence. In support of these predictions, mammalian cells spread on ECM substrates of varying affinities were challenged with bacteria coated with either invasin or MAbs directed against the $\alpha_5\beta_1$ integrin, and the efficiency of bacterial uptake was measured (TRAN VAN NHIEU and ISBERG 1993a). Bacteria coated with the high-affinity ligands were most efficiently internalized if the mammalian cells were plated on substrates having low affinities and were inefficiently internalized if the mammalian cells were plated on substrates containing high-affinity MAbs directed against $\alpha_5\beta_1$ integrin (TRAN VAN NHIEU and İSBERG 1993a).

A second explanation for the need for high-affinity binding by invasin proposes that uptake takes place via the "zipper model" (GRIFFIN et al. 1975). This postulates that internalization takes place by circumferential binding of mammalian cell receptors about the surface of the bacterium. More receptor is required for internalization than for simple adhesion of the bacterium, as the model requires a larger surface area of contact for uptake than for adhesion. In addition, the rate of dissociation of the receptor from the bacterial-encoded ligand must be rather slow, or else the zipper will fall apart before the internalization process is completed. As the rate of dissociation of the invasin-$\alpha_5\beta_1$ integrin complex corresponds to a $t_{1/2}=20$ min, this complex appears sufficiently stable to survive for the duration of the uptake process.

A final explanation for the need for high-affinity binding assumes the production of a signal within the mammalian cell. Receptor-ligand interactions that result in intracellular signaling events often require multimerization or clustering of receptors to send the proper signal (MADRENAS et al. 1995; SLOAN-LANCASTER et al. 1995), and invasin could cause such multimerization. If the production of this signal requires a high receptor density in a relatively small area of the mammalian cell surface, then high-affinity binding by invasin may be required to cause the receptor clustering of the proper density. Lower-affinity ligands may not allow a sufficient density of receptor to send the necessary signal.

7 Invasin Residues that Facilitate High-affinity Binding to Integrin Receptors

As described above, the region of invasin that exhibits sequence similarity to other proteins is limited to the amino terminal 700 amino acids. Surprisingly, the cell adhesion domain shows no extensive similarity to proteins that bind integrins, even though invasin appears to bind a site that is identical to that bound by other integrin ligands (TRAN VAN NHIEU and ISBERG 1991). Invasin must have important structural similarities to these other proteins not apparent from the primary sequence.

Invasin derivatives containing regions shorter than 192 amino acids fail to bind mammalian cells (ISBERG and LEONG 1990; L. SALTMAN and R. ISBERG, unpublished data). This indicates that multiple residues that contact receptor are dispersed throughout the carboxyl terminal, or that the entire cell-binding region is required to present critical residues. The importance of conformation is emphasized by two studies indicating that a disulfide bond within invasin is required for cellular adhesion, and that the carboxyl terminus of the protein is extremely sensitive to mutational changes.

Members of the family of proteins related to invasin and the intimins have a conserved pair of cysteines separated by 72–78 residues located in the extreme carboxyl terminal 80 amino acids. Evidence in favor of a disulfide bond between these two cysteines in invasin was obtained by analyzing random mutations that result in low cell adhesion but retain proper surface localization of invasin (LEONG and ISBERG 1993). Mutations causing the strongest defects in bacterial uptake harbored changes at either of the two carboxyl terminal cysteines located at residues 907 (C907) and 982 (C982), consistent with the presence of a disulfide bond between these two residues. Hybrid proteins having wild-type sequence are not alkylated by the sulfhydryl-specific reagent iodoacetic acid unless they have been previously reduced by dithiothreitol (LEONG and ISBERG 1993). Furthermore, partial proteolysis of the protein in the absence of reducing agent yields degradation products that are linked by disulfide bonds. Finally, mutations in either of the two cysteines cause hybrid proteins to form covalent dimers that can be converted to manomers in the presence of reducing agents, indicating that the single remaining cysteine participates in a disulfide bond that links the hybrid protein monomers.

Analysis of the extreme carboxyl terminus of invasin indicates either that the stereochemistry of the disulfide bond is critical for integrin recognition, or that residues important for cellular adhesion are located very close to one of the two cysteines. The addition of two amino acids (isoleucine and leucine) to the C-terminal residue of invasin totally eliminates cellular adhesion without eliminating the presence of the disulfide bond (ISBERG et al. 1993). The added residues could cause amino acids critical for adhesion to be rotated improperly relative to the disulfide bond, or they could interfere with receptor recognition of important side chains. Evidence for the latter model was obtained by isolating revertants that restored the ability to enter mammalian cells to the mutant harboring the added two amino acids. The revertants fell into two classes: those that inserted a stop codon at the same site as found in the wild-type gene, or those that inserted a glycine residue at the site of the wild-type stop codon. The presence of the glycine in the revertants probably removes interference of substrate recognition by offending side chains. As the extreme carboxyl terminus is only four residues downstream of the disulfide bond, the critical residues involved in integrin recognition obstructed by these mutations may be very close to one of the cysteines.

Further analysis of invasin mutants unable to promote uptake emphasized that residues critical for integrin recognition surround one of the two cysteines (LEONG et al. 1995). With the exception of mutations that appear to affect the disulfide bond or destabilize the protein, all mutations that affect recognition of integrins are in the vicinity of the C907 residue (LEONG et al. 1995). Subsequent site-directed oligonucleotide mutagenesis revealed that the only absolutely critical residue in this region appeared to be aspartate-911 (D911). Several residues near D911 in the primary sequence can be changed to alanine without affecting receptor recognition, whereas change of an aspartate to an alanine at this site (D911A) caused total loss of integrin binding (LEONG et al. 1995). Change of aspartate-911 to glutamate (D911E) also had profound effects, in spite of the conservative nature of this amino acid change. Bacteria harboring this mutation are unable to enter mammalian cells, although the purified invasin derivative containing the D911E change retained considerable receptor-binding activity (LEONG et al. 1995). This latter property is consistent with the model that uptake requires a higher-affinity interaction than does adhesion.

The sensitivity of D911 to conservative mutational changes argues for its functional similarity to important aspartate residues found in a number of integrin-binding proteins. In fibronectin, peptides containing the sequence arginine-glycine-aspartate (RGD) inhibit cellular adhesion, and change of the asp to a glu eliminates adhesiveness of either fibronectin or the RGD peptide (OBARA et al. 1988). The sequence containing aspartate 911 in the *Y. pseudotuberculosis* invasin protein is dissimilar to RGD, although the analogous sequence in the *Y. enterocolitica* protein is RTD (YOUNG et al. 1990). In VCAM-1, the only mutation that causes total loss of binding to integrin $\alpha_4\beta_1$ is the aspartate-34 (OSBORN et al. 1994; RENZ et al. 1994). Finally, fibrinogen also contains a critical aspartate in the sequence KQAGD (GINSBERG et al. 1992). There is no information regarding the reason for the importance of an aspartate, although several models hypothesize that it is important for recognition of a divalent cation known to be required for integrin adhesion to substrate (LOFTUS et al. 1990).

The final class of mutations that affect invasin recognition of receptor have mild affects and cluster about the residue aspartate-811 (D811; L. Saltman and R. Isberg, unpublished). They also cause instability of an otherwise wild-type protein under some circumstances, making their affects rather difficult to analyze. These important residues also argue for similarity between invasin and other integrin substrates. Many other integrin substrates have two domains involved in receptor recognition that are affected by mutational changes positioned approximately 80–100 amino acids apart on the primary sequence. Fibronectin, ICAM-1, and VCAM-1 all appear to consist of two immunoglobulin-like domains that have β-barrel-type structures affected by mutational changes in either domain. Perhaps these critical residues in invasin define regions of function similar to these other substrates.

8 Signaling and the Role of the Integrin β_1-chain Cytoplasmic Domain in Bacterial Uptake

Phosphorylated protein intermediates within the mammalian cell are critical for invasin-promoted uptake. *E. coli* strains that express the *Y. enterocolitica* invasin protein are poorly internalized by cultured cells in the presence of a variety of tyrosine protein kinase inhibitors (ROSENSHINE et al. 1992). The inhibitors appear to interfere with a step involved in the formation of the phagosome, as bacteria bind quite efficiently to mammalian cells in the presence of these drugs. A number of inhibitors show this effect, although the potency of the inhibition is dependent on the drug of choice. Staurosporine causes the strongest depression in uptake, perhaps due to lack of specificity of this particular reagent. Genistein and terphostin also inhibit uptake, although with lowered potency (ROSENSHINE et al. 1992).

Work on the *Yersinia* factor YopH supports the conclusion that tyrosine phosphorylation is important for uptake. As stated previously, YopH is deposited within the mammalian cell cytoplasm, with the result that bacterial uptake is inhibited (ROSQVIST et al. 1989, 1990). The plasmid-borne *YopH* protein shows high sequence similarity to mammalian protein tyrosine phosphatases that regulate a variety of intracellular signaling events (CLEMENS et al. 1991). Purified YopH protein also has potent phosphatase activity, and a point mutation in a conserved cysteine residue that disrupts activity of mammalian phosphatases eliminates YopH phosphatase activity (BLISKA et al. 1991). The point mutation results in lowered virulence of the microorganism, and bacteria harboring the mutation are relieved from YopH-dependent inhibition of uptake (BLISKA et al. 1991). Taken together, these results indicate that the inhibition of phagocytosis promoted by YopH is due to its removal of phosphates from proteins critical for uptake. It is important to note that the phosphatase activity appears to cause wholesale loss of phagocytic capability within the mammalian cell, as uptake via F_C receptors is similarly eliminated (BLISKA and BLACK 1995). Presumably, there is a phosphorylated intermediate common to a number of phagocytic processes that is the target of this enzyme. Several phosphorylated proteins have been isolated that are bound by a YopH catalytic mutant, and among these may be the critical target protein required for phagocytosis (BLISKA et al. 1992).

Signaling events may be required to initiate the critical cytoskeletal changes necessary for bacterial entry. Evidence has been presented for the fact that actin is found aggregated about phagosomes formed during invasin-promoted entry (YOUNG et al. 1992), and it is possible that this phenomenon is a result of signaling mediated by phosphorylated species. The presence of these actin cages could involve direct binding of the cytoskeleton to the cytoplasmic domain of the integrin β_1 chain, which is able to bind several cytoskeletal components, or result from interaction of the cytoskeleton with unknown membrane components that allow nucleation of actin filaments about the nascent phagosome.

There is surprising evidence that reduction in the affinity of the integrin β_1 chain for cytoskeletal components stimulates uptake (TRAN VAN NHIEU et al. 1995). Therefore, tight association of the integrin with the cytoskeleton may interfere with the ability of the host cell to internalize the microorganism. These results are derived from studies on mutations located in the integrin β_1-chain cytoplasmic domain that affect the ability of the receptor to participate in large complexes, called focal adhesions, containing a number of cytoskeletal components (BURRIDGE and FATH 1989). Focal adhesions occur in response to mammalian cell adherence to ECM components and consist of integrins binding extracellularly to substrate. Cytoskeletal components such as α-actinin, talin, p125[FAK], and vinculin accumulate at these sites in an array that appears anchored to actin stress fibers (HYNES 1987; KORNBERG et al. 1991; SCHALLER et al. 1992). The ability of the integrin to localize in such complexes is dependent on having an intact integrin β_1-chain cytoplasmic domain, as mutations located in this domain result in random surface localization of the receptor. Such mutants are deemed defective in interacting with cytoskeletal components. Point mutations that cause defects in focal adhesion formation are uniformly more efficient at promoting uptake than the wild-type receptor. This indicates that tight binding of the receptor to the cytoskeleton may immobilize the receptor and prevent it from clustering about the bacterium.

Mutations in the cytoplasmic domain of the integrin $\beta_1$chain that result in defective bacterial internalization are located within the tetrapeptide sequence NPKY. This sequence conforms to the NPXY sequence found in a number of receptors that undergo clathrin-dependent receptor-mediated endocytosis. Mutations in this sequence in the LDL receptor result in an inability to associate with clathrin coats and an inability to be internalized after binding ligand (CHEN et al. 1990). Clathrin and other endocytic factors may play a central role in integrin-mediated bacterial internalization. Large sheets of clathrin and the coat-associated protein AP-2 are readily found about the phagosomal surface during uptake, giving circumstantial support for their role in uptake. Antibody directed against clathrin or AP-2 loaded into mammalian cells also inhibit invasin-promoted uptake, indicating that these proteins probably play an important functional role in uptake (TRAN VAN NHIEU et al. 1995).

9 Animal Infection Models and Factors Associated with Cellular Uptake

As stated previously, pathogenic *Yersinia* species appear to maintain an extracellular niche deep within tissue sites, although intracellular bacteria are readily detected shortly after oral infection. Wolf-Watz and co-workers questioned whether factors responsible for bacterial uptake play a role in the establishment of infection in any tissue site, based on their analysis of *Y. pseudotuberculosis inv* mutants in an animal infection model (ROSQVIST et al. 1988b). They found that the 50% lethal dose (LD_{50}) of *Y. pseudotuberculosis* was unaffected by an insertion mutation in the *inv* gene, although mean time for death of the animals was delayed for the *inv* mutant after oral inoculation. The nature of this delay was not analyzed. Strangely, an *inv yadA* double mutant was more virulent than the wild-type strain, as the presence of these mutations resulted in a decrease in the LD_{50} by 2 orders of magnitude. As these two proteins are apparently nonfunctional in the closely related and more highly virulent *Y. pestis* species, the authors hypothesized that the presence of these two proteins is the reason why *Y. pseudotuberculosis* is a rather attenuated organism (ROSQVIST et al. 1988b). The hypervirulence of the double mutant is not supported by results from

Y. enterocolitica. Mutations in *yadA* in this bacterium result in decreased virulence (J. PEPE and V. MILLER, personal communication). As death is not the usual course of enteropathogenic *Yersinia* disease in human beings, but rather colonization of regional lymph nodes, LD_{50} measurements may not provide insight into the role of entry factors in disease.

Miller and co-workers performed a detailed analysis of colonization by *Y. enterocolitica inv* mutants (PEPE and MILLER 1993). As reported for *Y. pseudotuberculosis, inv* mutants had no affect on the LD_{50} of Balb/c mice. Localization in regional nodes by an *inv* insertion mutant, however, was profoundly affected; 24 h after initial oral infection, wild-type *Y. enterocolitica* strains showed heavy colonization of the Peyer's patches and mesenteric lymph nodes, with colonization 3–7 orders of magnitude higher than that found with the *inv* mutant. In spite of this fact, colonization of the liver and spleen by the *inv* mutant occurs with kinetics that are similar to those of the wild-type strain, which is apparently why the mutation has little affect on the LD_{50}. Data from Simonet are supportive of these results (SIMONET and FALKOW 1994). Mice that were immunized with vaccine strains of *S. typhimurium* harboring the *Y. pseudotuberculosis inv* gene produced antibody that apparently blocked function of invasin. On subsequent challenge with virulent *Y. pseudotuberculosis,* colonization of regional lymph nodes was inhibited in the immunized mice. Expression of invasin appears to be required for efficient rapid localization in regional lymph nodes, the usual site of colonization in a human infection. Furthermore, the fact that *inv* mutants were able to penetrate into deep tissues without rapid localization into regional lymph nodes indicates that there exists a second invasin-independent pathway responsible for translocation across the intestinal mucosa. As localized damage to the intestinal epithelial layer is possible, perhaps microdamage in the intestine provides a portal of entry for the *inv* mutants.

10 Conclusions and Future Prospects

A number of important lessons have been learned from studies on *Yersinia* entry. This paradigm emphasizes the fact that a bacterial pathogen can have multiple strategies for promoting internalization into mammalian cells, and it would not be surprising if other pathogens had this property. Even more striking is the fact that although invasin is required for efficient translocation from the lumen of the intestine into the Peyer's patch, there appears to be an alternate way for the microorganism to penetrate the intestinal mucosal in the absence of invasin. The route of transit that allows the microorganism to appear in the liver and spleen after initial ingestion is somewhat mysterious, however, as there is little evidence of colonization in intermediate sites between the intestine and deeper tissues. This result has some parallels with *Salmonella typhimurium* mutants defective for uptake into cultured cells. Such mutants are lowered for their LD_{50} in mouse infection models, but they still kill mice at doses similar to those seen with wild-type *Y. enterocolitica* (GALAN and CURTISS 1989). Presumably, these mutants also have an alternate strategy for translocation across the intestinal epithelium.

Invasin is an important example of a bacterial ligand able to bind mammalian integrin receptors. A number of bacterial, viral, and fungal pathogens encode ligands that allow adhesion to integrins (ISBERG and TRAN VAN NHIEU 1994b). Many of these pathogens, such as adenovirus (WICKHAM et al. 1993) and the enteropathogenic *Yersinia*, are internalized into the epithelium, in spite of the fact that integrin receptors are presumably localized on the basal end of target cells not readily accessible to the microorganisms (HYNES 1992; RUOSLAHTI and PIERSCHBACHER 1987). The apparent explanation for this phenomenon, at least in the example of invasin binding, is that the vast majority of cells in the epithelium do not bind the microorganism. Rather, a subset of cells act as targets for binding and subsequent translocation of the microorganism (GRUTZKAU at al. 1990). These cells are M cells, and they apparently have integrin receptors promiscuously distributed about their surface (A. MARRA and R. ISBERG, unpublished data). This affords sites of translocation for microorganisms bearing integrin ligands, or portals that allow spread of viruses laterally through the epithelium after initially entering a subset of cells.

The high-affinity model for invasin-promoted uptake also has parallels in other systems which may provide insight into the mechanism of phagocytosis. The ability of T cells to respond to presentation of antigen by major histocompatibility complex (MHC) molecules also appears to be controlled by affinity. The ability of T cells to respond to peptides associated with MHC molecules on antigen-presenting cells appears to be dependent upon the affinity of the T-cell receptor for the peptide antigen (MADRENAS et al. 1995). Peptides that have the highest affinity for the T-cell receptor cause clustering of receptor and efficient intra-cellular signaling. Peptides of lower affinity are able to ligate T-cell receptors but are unable to stimulate signaling within T cells. This is quite similar to the contrast between high-affinity ligands for integrins that promote bacterial uptake and low-affinity ligands that are limited to extracellular adhesion, which may be explained by different abilities to send the appropriate signals for internalization. Identification of the cytosolic machinery that is sensitive to this signaling and determination of how these signals are transduced will be major challenges in the future.

References

Akiyama SK, Yamada KM (1985) The interaction of plasma fibronectin with fibroblastic cells in suspension. J Biol Chem 260: 4492–5000

Allaoui A, Sansonetti PJ, Parsot C (1993) MxiD, an outer membrane protein necessary for the secretion of the *Shigella flexneri* Ipa invasins. Mol Microbiol 7: 59–68

Andersen JK, Sorensen R, Glensbjerg M (1991) Aspects of the epidemiology of *Yersinia enterocolitica*: a review. Int J Food Microbiol 13: 231–237

Argraves WS, Pytela R, Suzuki S, Millan JL, Pierschbacher MD, Ruoslahti E (1986) cDNA sequences from the alpha subunit of the fibronectin receptor predict a transmembrane domain and a short cytoplasmic peptide. J Biol Chem 261: 12922–12924

Beer KB, Miller VL (1992) Amino acid substitutions in naturally occurring variants of *ail* result in altered invasion activity. J Bacteriol 174: 1360–1369

Bliska JB, Black DS (1995) Inhibition of the Fc receptormediated oxidative burst in macrophages by the *Yersinia pseudotuberculosis* tyrosine phosphatase. Infect Immun 63: 681–685

Bliska JB, Falkow S (1992) Bacterial resistance to complement killing mediated by the Ail protein of *Yersinia enterocolitica*. Proc Natl Acad Sci USA 89: 3561–3565

Bliska JB, Guan KL, Dixon JE, Falkow S (1991) Tyrosine phosphate hydrolysis of host proteins by an essential *Yersinia* virulence determinant. Proc Nat Acad Sci USA 88: 1187–1191

Bliska JB, Clemens JC, Dixon JE, Falkow S (1992) The *Yersinia* tyrosine phosphatase: specificity of a bacterial virulence determinant for phosphoproteins in the J774A.1 macrophage. J Exp Med 176: 1625–1630

Bliska JB, Copass MC, Falkow S (1993) The *Yersinia pseudotuberculosis* adhesin YadA mediates intimate bacterial attachment to and entry into HEp-2 cells. Infect Immun 61: 3914–3921

Bovallius A, Nilsson G (1975) Ingestion and survival of *Yersinia pseudotuberculosis* in HeLa cells. Can J Microbiol 7: 1997–2007

Burridge K, Fath K (1989) Focal contacts transmembrane links between the extracellular matrix and the cytoskeleton. Bioessays 10: 104–108

Carter PB (1975) Pathogenecity of *Yersinia enterocolitica* for mice. Infect Immun 11: 164–70

Chen WJ, Goldstein JL, Brown MS (1990) NPXY, a sequence often found in cytoplasmic tails, is required for coated pit-mediated internalization of the low-density lipoprotein receptor. J Biol Chem 265: 3116–3123

China B, Sory MP, Nguyen BT, De Bruyere M, Cornelis GR (1993) Role of the YadA protein in prevention of opsonization of *Yersinia enterocolitica* by C3b molecules. Infect Immun 61: 3129–3136

China B, N'Guyen BT, De Bruyere M, Cornelis GR (1994) Role of YadA in resistance of *Yersinia enterocolitica* to phagocytosis by human polymorphonuclear leukocytes. Infect Immun 62: 1275–1281

Clemens JC, Guan K, Bliska JB, Falkow S, Dixon JE (1991) Microbial pathogenesis and tyrosine dephosphorylation: surprising 'bedfellows'. Mol Microbiol 5: 2617–2620

Cornelis GR, Biot T, Lambert de Rouvroit C, Michiels T, Mulder B, Sluiters C, Sory MP, Van BM, Vanooteghem JC (1989) The *Yersinia* Yop regulon. Mol Microbiol 3: 1455–9145

Devenish JA, Schiemann DA (1981) HeLa cell infection by *Yersinia enterocolitica*: evidence for lack of intracellular multiplication and development of a new procedure for quantitative expression of infectivity. Infect Immun 32: 48–55

Donnenberg MS, Yu J, Kaper JB (1993) A second chromosomal gene necessary for intimate attachment of enteropathogenic Escherichia coli to epithelial cells 175: 4670–4680

Emody L, Heesemann J, Wolf-Watz H, Skurnik M, Kapperud G, O'Toole P, Wadstrom T (1989) Binding to collagen by *Yersinia enterocolitica* and *Yersinia pseudotuberculosis*: evidence for YopA-mediated and chromosomally encoded mechanisms. J Bacteriol 171: 6674–6679

Ennis E, Isberg RR, Shimizu Y (1993) VLA4-dependent adhesion and costimulation of resting human T cells by the bacterial β1 integrin ligand invasin. J Exp Med 177: 207–212

Fields KA, Plano GV, Straley SC (1994) A low-Ca2+ response (LCR) secretion (Ysc) locus lies within the lcrB region of the LCR plasmid in *Yersinia pestis*. J Bacteriol 176: 569–57

Finlay BB, Falkow S (1988) Comparison of the invasion strategies used by *Salmonella choleraesuis*, *Shigella flexneri* and *Yersinia* enterocolitica to enter cultured animal cells: endosome acidification is not required for bacterial invasion or intracellular replication. Biochimie 70: 1089–1099

Flock JI, Froman G, Jonsson K, Guss B, Signas C, Nilsson B, Raucci G, Hook M, Wadstrom T, Lindberg M (1987) Cloning and expression of a gene for a fibronectin-binding protein from Staphylococcus aureus. EMBO J 6: 2351–2357

Fowler JM, Brubaker RR (1994) Physiological basis of the low calcium response in *Yersinia pestis*. Infect Immun 62: 5234–5241

Frankel G, Candy DC, Everest P, Dougan G (1994) Characterization of the C-terminal domains of intimin-like proteins of enteropathogenic and enterohemorrhagic *Escherichia coli, Citrobacter freundii*, and *Hafnia alvei*. Infect Immun 62: 1835–1842

Fukai K, Maruyama T (1979) Histopathological studies on experimental *Yersinia enterocolitica* infection in animals. Contrib Microbiol Immunol 5: 310–316

Fukushima H (1991) Susceptibility of wild mice to *Yersinia pseudotuberculosis* and *Yersinia entero-colitica*. Int J Med Microbiol 275: 530–40

Fukushima H, Gomyoda, M, Kaneko, S (1990) Mice and moles inhabiting mountainous areas of Shimane Peninsula as sources of infection with *Yersinia pseudotuberculosis*. J Clin Microbiol 28: 2448–55

Galan JE, Curtiss R (1989) Cloning and molecular characterization of genes whose products allow *Salmonella typhimurium* to penetrate tissue culture cells. Proc Natl Acad Sci USA 86: 6383–6387

Galyov EE, Hakansson S, Forsberg A, Wolf-Watz H (1993) A secreted protein kinase of *Yersinia pseudotuberculosis* is an indispensable virulence determinant. Nature 361: 730–732

Ginsberg MH, Du X, Plow EF (1992) Inside-out integrin signalling. Curr Opin Cell Biol 4: 766–771

Griffin FMJ, Griffin JA, Leider JE, Silverstein SC (1975) Studies on the mechanism of phagocytosis. I. Requirements for circumferential attachment of particle-bound ligands to specific receptors on the macrophage plasma membrane. J Exp Med 142: 1268–1282

Grutzkau A, Hanski C, Hahn H, Riecken EO (1990) Involvement of M cells in the bacterial invasion of Peyer's patches: a common mechanism shared by *Yersinia enterocolitica* and other enteroinvasive bacteria. Gut 31: 1011–5

Guan JL, Trevithick JE, Hynes RO (1991) Fibronectin/integrin interaction induces tyrosine phosphorylation of a 120-kDa protein. Cell Regul 2: 951–964

Haddix PL, Straley SC (1992) Structure and regulation of the *Yersinia pestis YscBCDEF* operon J Bacteriol 174: 4820–4828

Hanski C, Kutschka U, Schmoranzer HP, Naumann M, Stallmach A, Hahn H, Menge H, Riecken EO (1989a) Immunohistochemical and electron microscopic study of interaction of *Yersinia enterocolitica* serotype 08 with intestinal mucosa during experimental enteritis. Infect Immun 57: 673–8

Hanski C, Naumann M, Hahn H, Riecken EO (1989b) Determinants of invasion and survival of *Yersinia enterocolitica* in intestinal tissue: an in vivo study. Med Microbiol Immunol (Berl) 178: 289–96

Heesemann J, Gaede K, Autenrieth IB (1993) Experimental *Yersinia enterocolitica* infection in rodents: a model for human yersiniosis. APMIS 101: 417–429

Heffernan EJ, Wu L, Louie J, Okamoto S, Fierer J, Guiney DG (1994) Specificity of the complement resistance and cell association phenotypes encoded by the outer membrane protein genes *rck* from *Salmonella typhimurium* and *ail* from *Yersinia enterocolitica*. Infect Immun 62: 5183–5186

Heffernan EJ, Harwood J, Fierer J, Guiney D (1992) The *Salmonella typhimurium* virulence plasmid complement resistance gene *rck* is homologous to a family of virulence-related outer membrane protein genes, including *pagC* and *ail*. J Bacteriol 174: 1360–1369

Hermanowski-Vosatka A, Van Strijp JA, Swiggard WJ, Wright SD (1992) Integrin modulating factor-1: a lipid that alters the function of leukocyte integrins. Cell 68: 341–352

Hook M, Switalski L, Wastrom T, Lindberg M (1989) Interactions of pathogenic microorganisms with fibronectin. In: Mosher DF (ed) Fibronectin. Academic, San Diego, pp 295–308

Horwitz A, Duggan K, Buck C, Beckerle MC, Burridge K (1986) Interaction of plasma membrane fibronectin receptor with talin—a transmembrane linkage. Proc Natl Acad Sci USA 83: 6470–6474

Huang HC, He SY, Bauer DW, Collmer A (1992) The *Pseudomonas syringae* pv. syringae 61 *hrpH* product, an envelope protein required for elicitation of the hypersensitive response in plants. J Bacteriol 174: 6878–6885

Hynes RO (1987) Integrins: a family of cell surface receptors. Cell 48: 549–554

Hynes RO (1992) Integrins: versatility, modulation, and signaling in cell adhesion. Cell 69: 11–25

Isberg RR (1989) Two determinants for *Yersinia pseudotuberculosis* cell binding and penetration detected in the absence of invasin. Infect Immun 57: 1998–2005

Isberg RR, Falkow S (1985) A single genetic locus encoded by *Yersinia pseudotuberculosis* permits invasion of cultured animal cells by *Escherichia coli* K-12. Nature 317: 262–264

Isberg RR, Leong JM (1988) Cultured mammalian cells attach to the invasin protein of *Yersinia pseudotuberculosis*. Proc Natl Acad Sci USA 85: 6682–6686

Isberg RR, Leong JM (1990) Multiple β1 chain integrins are receptors for invasin, a protein that promotes bacterial penetration into mammalian cells. Cell 60: 861–871

Isberg RR, Voorhis DL, Falkow S (1987) Identification of invasin: a protein that allows enteric bacteria to penetrate cultured mammalian cells. Cell 50: 769–778

Isberg RR, Swain A, Falkow S (1988) Analysis of expression and thermoregulation of the *Yersinia pseudotuberculosis inv* gene with hybrid proteins. Infect Immun 56: 2133–2138

Isberg RR, Tran Van Nhieu G (1994a) Two mammalian cell internalization strategies used by pathogenic bacteria. Annu Rev Genet 28: 395–422

Isberg RR, Tran Van Nhieu G (1994b) Binding and internalization of microorganisms by integrin receptors. Trends Microbiol 2: 10–14

Isberg RR, Yang Y, Voorhis DL (1993) Residues added to the Carboxyl terminus of the *Yersinia pseudotuberculosis* invasin protein interfere with recognition by integrin receptors. J Biol Chem 268: 15840–15846

Jerse AE, Yu J, Tall BD, Kaper JB (1990) A genetic locus of enteropathogenic *Escherichia coli* necessary for the production of attaching and effacing lesions on tissue culture cells. Proc Natl Acad Sci USA 87: 7839–7843

Kornberg LJ, Earp HS, Turner CE, Prockop C, Juliano RL (1991) Signal transduction by integrins: increased protein tyrosine phosphorylation caused by clustering of β1 integrins. Proc Natl Acad Sci USA 88: 8392–8396

Kreusch A, Schulz GE (1994) Refined structure of the porin from *Rhodopseudomonas blastica*. Comparison with the porin from *Rhodobacter capsulatus*. J Mol Biol 243: 891–905

Lam SC, Plow EF, D'Souza SE, Cheresh DA, Frelinger AL, Ginsberg MH (1990) Isolation and characterization of a platelet membrane protein related to the vitronectin receptor. J Biol Chem 265: 3440–3446

Leong JM, Isberg RR (1993) The formation of a disulfide bond is required for efficient integrin binding by invasin. J Biol Chem 268: 20524–20532

Leong JM, Fournier RS, Isberg RR (1990) Identification of the integrin binding domain of the *Yersinia pseudotuberculosis* invasin protein. EMBO J 9: 1979–1989

Leong JM, Morrissey, PE, Marra A, Isberg RR (1995) An aspartate residue of the *Yersinia pseudotuberculosis* invasin protein that is critical for integrin binding. EMBO J 14: 422–431

Lindler LE, Tall BD (1993) *Yersinia pestis* pH 6 antigen forms fimbriae and is induced by intracellular association with macrophages. Mol Microbiol 8: 311–24

Loftus JC, O'Toole TE, Plow EF, Glass A, Frelinger AL, Ginsberg MH (1990) A β3 integrin mutation abolishes ligand binding and alters divalent cation-dependent conformation. Sciencce 249: 915–918

Madrenas J, Wange RL, Wang JL, Isakov N, Samelson LE, Germain RN (1995) Zeta phosphorylation without ZAP-70 activation induced by TCR antagonists or partial agonists. Science 267: 515–518

Maki IO, Viljanen MK, Tiitinen S, Toivanen P, Granfors K (1991) Antibodies to arthritis-associated microbes in inflammatory joint diseases. Rheumatol Int 10: 231–4

Maki IO, Pulz M, Heesemann J, Lahesmaa R, Saario R, Toivanen A, Granfors K (1992) Antibody response against 26 and 46 kilodalton released proteins of *Yersinia* in *Yersinia*-triggered reactive arthritis. Ann Rheum Dis 51: 1247–1249

Mandell GL (1973) Interaction of intraleukocytic bacteria and antibiotics. J Clin Invest 52: 1673–1679

Mecsas J, Welch R, Erickson JW, Gross CA (1995) Identification and characterization of an outer membrane protein, ompX, in Escherichia coli that is homologous to a family of outer membrane proteins including Ail of Yersinia enterocolitica. J Bacteriol 177: 799–804

Merilahti PR, Lahesmaa R, Granfors K, Gripenberg-Lerche C, Toivanen P (1991) Risk of *Yersinia* infection among butchers. Scand J Infect Dis 23: 55–61

Michiels T, Vanooteghem JC, Lambert de Rouvroit C, China B, Gustin A, Boudry P, Cornelis GR (1991) Analysis of *virC*, an operon involved in the secretion of Yop proteins by *Yersinia enterocolitica*. J Bacteriol 173: 4994–5009

Miller VL, Falkow S (1988) Evidence for two genetic loci from *Yersinia enterocolitica* that can promote invasion of epithelial cells. Infect Immun 56: 1242–1248

Miller VL, Farmer JJ, Hill WE, Falkow S (1989) The ail locus is found uniquely in *Yersinia enterocolitica* serotypes commonly associated with disease. Infect Immun 57: 121–31

Miller VL, Bliska JB, Falkow S (1990) Nucleotide sequence of the *Yersinia enterocolitica ail* gene and characterization of the Ail protein product. J Bact eriol 172: 1062–1069

Munn AL, Reeves P (1985) High level synthesis of the phage λ outer membrane protein from the cloned lom gene. Gene 38: 253–258

Niewiarowski S, McLane MA, Kloczewiak M, Stewart GJ (1994) Disintegrins and other naturally occurring antagonists of platelet fibrinogen receptors. Semin Hematol 31: 289–300

O'Loughlin EV, Gall DG, Pai CH (1990) *Yersinia enterocolitica*: mechanisms of microbial pathogenesis and pathophysiology ofdiarrhoea. J Gastroent erol Hepatol 5: 173–179

Obara M, Kang MS, Yamada KM (1988) Site-directed mutagenesis of the cell-binding domain of human fibronectin: separable, synergistic sites mediate adhesive function. Cell 53: 649–657

Osborn L, Vassallo C, Browning BG, Tizard R, Haskard DO, Benjamin CD, Dougas I, Kirchhausen T (1994) Arrangement of domains, and amino acid residues required for binding of vascular cell adhesion molecule-1 to its counter-receptor VLA-4 ($\alpha_4\beta_1$). J Cell Biol 124: 601–8

Otey CA, Pavalko FM, Bruuidge K (1990) An interaction between α-actinin and the β1 integrin subunit in vitro. J Cell Biol 111: 721–729

Pepe J, Miller VL (1993) *Yersinia enterocolitica* invasin: a primary role in the initiation of infection. Proc Natl Acad Sci USA 90: 6373–6377

Pierson DE (1994) Mutations affecting lipopolysaccharide enhance *ail*-mediated entry of *Yersinia enterocolitica* into mammalian cells. J Bacteriol 176: 4043–4051

Pierson DE, Falkow S (1993) The *ail* gene of *Yersinia enterocolitica* has a role in the ability of the organism to survive serum killing. Infect Immun 61: 1846–1852

Pulkkinen WS, Miller SI (1991) A *Salmonella typhimurium* virulence protein is similar to a *Yersinia enterocolitica* invasin protein and a bacteriophage lambda outer membrane protein. J Bacteriol 173: 86–93

Rankin S, Isberg RR, Leong JM (1992) The integrin-binding domain of invasin is sufficient to allow bacterial entry into mammalian cells. Infect Immun 60: 3909–3912

Renz ME, Chiu HH, Jones S, Fox J, Kim KJ, Presta LG, Fong S (1994) Structural requirements for adhesion of soluble recombinant murine vascular cell adhesion molecule-1 to α4β1. J Cell Biol 125: 1395–1406

Rimpilainen M, Forsberg A, Wolf WH (1992) A novel protein, LcrQ, involved in the low-calcium response of *Yersinia pseudotuberculosis* shows extensive homology to YopH. J Bacteriol 174: 3355–63

Rosenshine I, Duronio V, Finley BB (1992) Tyrosine protein kinase inhibitors block invasin-promoted bacterial uptake by epithelial cells. Infect Immun 60: 2211–2217

Rosqvist R, Bolin I, Wolf-Watz H (1988a) Inhibition of phagocytosis in *Yersinia pseudotuberculosis:* a virulence plasmid-encoded ability involving the Yop2b protein. Infect Immun 56: 2139–43

Rosqvist R, Skurnik M, Wolf-Watz H (1988b) Increased virulence of *Yersinia pseudotuberculosis* by two independent mutations.

Rosqvist R, Forsberg A, Rimpilainen M, Bergman T, Wolf WH (1990) The cytotoxic protein YopE of *Yersinia* obstructs the primary host defence. Mol Microbiol 4: 657–67

Rosqvist R, Forsberg A, Wolf-Watz H (1991) Intracellular targeting of the *Yersinia* YopE cytotoxin in mammalian cells induces actin microfilament disruption. Infect Immun 59: 4562–4569 Nature 334: 522–525

Ruoslahti E, Pierschbacher MD (1987) New perspectives in cell adhesion: RGD and integrins. Science 238: 491–497

Saario R, Leino R, Lahesmaa R, Granfors K, Toivanen A (1992) Function of terminal ileum in patients with *Yersinia*-triggered reactive arthritis. J Intern Med 232: 73–76

Schaller MD, Borgman CA, Cobb BS, Vines RR, Reynolds AB, Parsons JT (1992) pp125[FAK], a structurally distinctive protein-tyrosine kinase associated with focal adhesions. Proc Natl Acad Sci USA 89: 5192–5196

Schauer DB, Falkow S (1993) The *eae* gene of *Citrobacter freundii* biotype 4280 is necessary for colonization in transmissible murine colonic hyperplasia. Infect Immun 61: 4654–4661

Schulze-Koops H, Burkhardt H, Heesemann J, Kirsch T, Swoboda B, Bull C, Goodman S, Emmrich F (1993) Outer membrane protein YadA of enteropathogenic yersiniae mediates specific binding to cellular but not plasma fibronectin. Infect Immun 61: 2513–2519

Schwartz MA, Lechene C, Ingber DE (1991) Insoluble fibronectin activates the Na/H antiporter by clustering and immobilizing integrin α5 β1, independent of cell shape. Proc Natl Acad Sci USA 88: 7849–7853

Simonet M, Falkow S (1994) Immunization with live *aroA* recombinant *Salmonella typhimurium* producing invasin inhibits intestinal translocation of *Yersinia pseudotuberculosis*. Infect Immun 62: 863–867

Simonet M, Richard S, Berche P (1992) Electron microscopic evidence for in vivo extracellular localization of *Yersinia pseudotuberculosis* harboring the pYV plasmid. Infect Immun 60: 366–373

Skurnik M, Wolf-Watz H (1989) Analysis of the *yopA* gene encoding the Yop1 virulence determinants of *Yersinia* spp. J Bacteriol 171: 6674–6679

Sloan-Lancaster J, Shaw AS, Rothbard JB, Allen PM (1995) Partial T cell signaling: altered phospho-zeta and lack of zap70 recruitment in APL-induced T cell anergy. Cell 79: 913–922

Stahlberg TH, Tertti R, Wolf-Watz H, Granfors K, Toivanen A (1987) Antibody response in *Yersinia pseudotuberculosis* III infection: analysis of an outbreak. J Infect Dis 156: 388–391

Stoorvogel J, van Bussel MJ, Tommassen J, van de Klundert JA (1991) Molecular characterization of an Enterbacter cloacae outer membrane protein (OmpX). J Bacteriol 173: 156–160

Straley SC, Bowmer WS (1986) Virulence genes regulated at the transcriptional level by Ca^{2+} in *Yersinia pestis* include structural genes for outer membrane proteins. Infect Immun 51: 445–454

Straley SC, Plano GV, Skrzypek E, Haddix PL, Fields KA (1993) Regulation by Ca2+ in the *Yersinia* low-Ca2+ response. Mol Microbiol 8: 1005–1010

Takada Y, Ylanne J, Mandelman D, Puzon W, Ginsberg MH (1992) A point mutation of integrin β1 subunit blocks binding of α5 β1 to fibronectin and invasin but not recruitment to adhesion plaques. J Cell Biol 119: 913–921

Toivanen P, Toivanen A (1994) Does *Yersinsia* induce autoimmunity? Int Arch Allergy Immunol 104: 107–111

Toivanen A, Yli KT, Luukkainen R, Merilahti PR, Granfors K, Seppala J (1993) Effect of antimicrobial treatment on chronic reactive arthritis. Clin Exp Rheumatol 11: 301–307

Tomer Y, Davies TF (1993) Infection, thyroid disease, and autoimmunity. Endocr Rev 14: 107–120

Tran Van Nhieu G, Isberg RR (1991) The *Yersinia pseudotuberculosis* invasin protein and human fibronectin bind to mutually exclusive sites on the α5β1 integrin receptor. J Biol Chem 266: 24367–24375

Tran Van Nhieu G, Isberg RR (1993a) Affinity and receptor density are primary determinants of β1 chain integrin-mediated bacterial internalization. EMBO J 12: 1887–1895

Tran Van Nhieu G, Isberg RR (1993b) Monoclonal antibodies as ligands promoting integrin-mediated bacterial internalization by cultured cells. ImmunoMethods 2: 71–77

Tran Van Nhieu G, Reszka A, Horwitz AF, Isberg RR (1995) A determinant of receptor-mediated endocytosis is required for integrin-promoted bacterial internalization J Biol Chem (Submitted for publication)

Tripoli LC, Brouillette DE, Nicholas JJ, Van TD (1990) Disseminated *Yersinia enterocolitica*. Case report and review of the literature. J Clin Gastroenterol 12: 85–89

Wenzel BE, Heesemann J, Heufelder A, Franke TF, Grammerstorf, S, Stemerowicz, R, Hopf, U (1991) Enteropathogenic *Yersinia enterocolitica* and organ-specific autoimmune diseases in man. Contrib Microbiol Immunol 12: 80–88

Wickham TJ, Mathias P, Cheresh DA, Nemerow GR (1993) Integrins αvβ3 and αvβ5 promote adenovirus internalization but not virus attachment. Cell 73: 309–319

Yang Y, Isberg RR (1993) Cellular internalization in the absence of invasin expression is promoted by the *Yersinia pseudotuberculosis* yadA product. Infect Immun 61: 3907–3913

Young VB, Falkow S, Schoolnik GK (1992) The invasin protein of *Yersinia enterocolitica*: internalization of invasin-bearing bacteria by eukaryotic cells is associated with reorganization of the cytoskeleton. J Cell Biol 116: 197–207

Young VB, Miller VL, Falkow S, Schoolnik GK (1990) Sequence, localization and function of the invasin protein of *Yersinia enterocolitica*. Mol Microbiol 4: 1119–28

Yu J, Kaper JB (1992) Cloning and characterization of the *eae* gene of enterohaemorrhagic *Escherichia coli* O157: H7. Mol Microbiol 6: 411–417

Invasion and the Pathogenesis
of *Shigella* Infections

C. Parsot and P.J. Sansonetti

1 Introduction

Bacillary dysentery, or shigellosis, is caused by penetration of *Shigella* spp into the intestinal mucosa of the colon, where degeneration of the epithelium and a strong inflammatory reaction indicate sites of infection (LaBrec et al. 1964). Clinical signs of shigellosis range from mild diarrhea to severe dysentery with blood, mucus, and pus in the stool. Epidemiological studies indicate that Shigella are transmitted by the fecal-oral route and sometimes by contaminated food (Wharton et al. 1990). *Shigella* are highly infectious organisms for human beings, since only a few hundred bacteria administrated orally caused disease in 50% of volunteers (DuPont et al. 1989).

Unité de Pathogénie Microbienne Moléculaire, Unité INSEM 389, Département de Bactériologie et de Mycologie, Institut Pasteur, 25 rue du Dr Roux, 75724 Paris Cedex 15, France

The genus *Shigella* is divided into four species, *S. boydii*, *S. dysenteriae*, *S. flexneri*, and *S. sonnei*. The cellular biology and genetics of entry and intercellular dissemination have been investigated using mainly *S. flexneri*, but most conclusions derived from these studies probably apply to other *Shigella* species, as well as to enteroinvasive *Escherichia coli* (EIEC) that cause a dysentery-like syndrome similar to shigellosis (DuPONT et al. 1971). Most of our knowledge about the pathogenesis of the disease is derived from studies using experimentally infected monkeys (TAKEUCHY et al. 1968). Injection of bacteria into rabbit ligated ileal loop, which elicits fluid accumulation and mucosal destruction (GOTS et al. 1974), and infection of the corneal epithelium of guinea pigs, which provokes keratoconjunctivitis (SERENY 1957), are also used to assess *Shigella* virulence.

The various aspects of *Shigella* pathogenicity have recently been reviewed (HALE 1991; SANSONETTI 1991; PARSOT 1994; see also SANSONETTI 1992). The current review will focus on the genetics and biology of entry and intercellular dissemination, as they have been studied on epithelial cells in vitro, and on recent data on the mechanism of invasion in vivo. Moreover, without underestimating the importance of chromosomal genes with respect to the invasion ability of *Shigella* (SANSONETTI et al. 1983b; OKAMURA et al. 1983; OKADA et al. 1991a,b; BERNARDINI et al. 1993), we shall focus the discussion of the genetics of entry and dissemination on genes which are carried by the virulence plasmid.

Infection of polarized cells which were differentiated from the human colonic epithelial cell line Caco-2 indicated that *S. flexneri* enters through the basolateral pole rather than the apical pole of epithelial cells (MOUNIER et al. 1992). In vitro, infection of epithelial cells by *Shigella* is a multistep process involving (a) adhesion of bacteria to the cell, (b) entry by induced endocytosis, (c) escape from the endosome, which completes the process of entry, (d) intracellular multiplication, (e) polymerization of actin filaments and reorganization of these filaments at one pole of the dividing bacteria to generate a movement leading to the formation of protrusions, and (f) lysis of the two cellular membranes surrounding bacteria once protrusions have entered into adjacent cells, which completes the process of intercellular dissemination (Fig. 1).

2 Entry into Epithelial Cells

2.1 Biology of Entry

Since *Shigella* invade epithelial and nonepithelial cells in vitro (GERBER and WATKINS 1961; LABREC et al. 1964), the entry process has been studied mainly with cultured cell lines. Following adhesion to the cell, a step that is quite elusive in the case of *Shigella*, bacteria are internalized by epithelial cells in a process similar to phagocytosis, in that it requires actin polymerization and myosin accumulation at

Entry   Dissemination

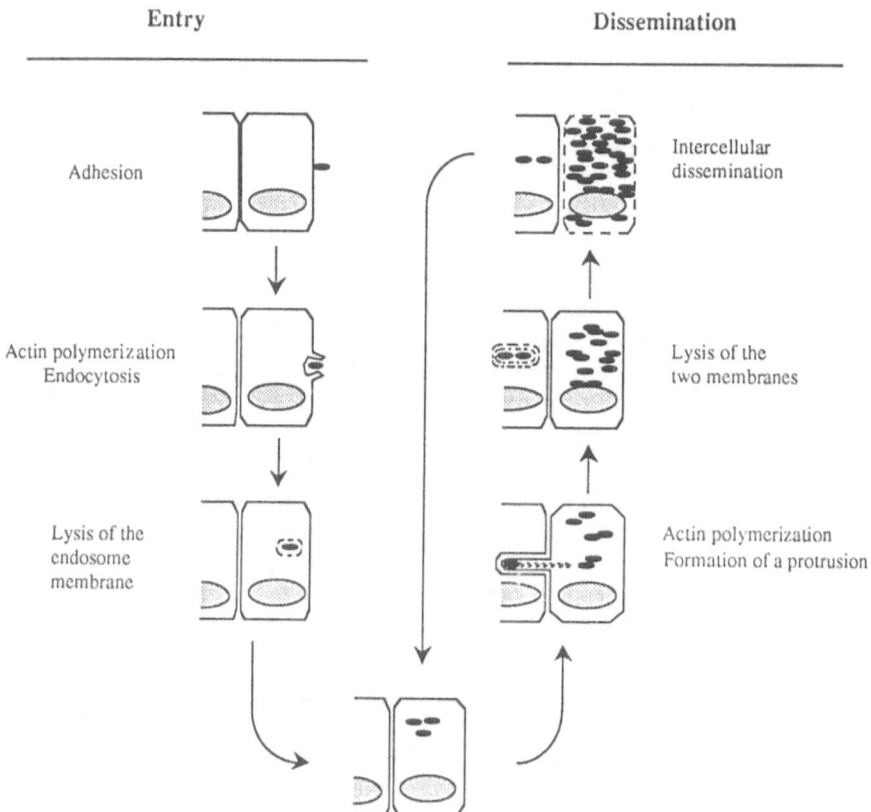

Adhesion

Actin polymerization
Endocytosis

Lysis of the
endosome
membrane

Intercellular
dissemination

Lysis of the
two membranes

Actin polymerization
Formation of a protrusion

Fig. 1. Entry and dissemination of *Shigella* in epithelial cells in vitro

the site of entry (HALE et al. 1979; CLERC and SANSONETTI 1987). Within a few minutes after entry, *Shigella* lyse the endosome membrane and gain access to the cytoplasm of the cell, where they multiply with a generation time of about 40 min (SANSONETTI et al. 1986). Invasion by *S. flexneri* of a murine macrophage cell line, J774, resulted in a rapid killing of the host cell (SANSONETTI and MOUNIER 1987). Wild-type *Shigella* induced apoptosis, i.e., programmed cell death, in infected macrophages, both J774 and mouse peritoneal macrophages (ZYCHLINSKY et al. 1992).

Recent work has led to a better characterization of the cellular basis of *Shigella* entry. Like *Salmonella*, *Shigella* do not follow the paradigm of *Yersinia pseudotuberculosis* invasion, in which the microorganism enters epithelial cells by establishing sequential high-affinity ligand-receptor interactions; in this case, the surface-expressed invasin binds to host-cell β1 integrins and promotes entry through a "zippering-like process" (ISBERG and LEONG 1990). *Shigella* and *Salmonella* appear to enter via a process of macropinocytosis; upon contact of bacteria with the epithelial cell surface, massive rearrangements of the cytoskeleton are induced which cause localized membrane ruffles, achieving bacterial uptake (FRANCIS et al. 1993; ADAM et al. 1995). These cytoskeletal rearrangements are characterized by induction or recruitment of numerous actin nucleation foci on the inner face of the cytoplasmic membrane, at the site of interaction between bacterial and cell surfaces (ADAM et al. 1995).

Little is currently known concerning the cellular signals inducing this early process. However, recent evidence indicates that the protooncogene pp60$^{c\text{-}src}$ is recruited and activated at the entry site, thus leading, among other phenotypes, to tyrosine-phosphorylation of cortactin, a major substrate for c-src (DEHIO et al. 1995). Cortactin, a 80-kD actin-binding protein (WU et al. 1991; WU and PARSONS 1993), might play a role in cytoskeletal rearrangements at this early stage. Future work will address the interaction that occurs between the *Shigella* invasion proteins (see below) and activation of the protooncogene. Ongoing experiments also indicate that elongation of actin filaments from the nucleation foci occurs underneath the cytoplasmic membrane and is under the control of the small GTPase rho (ADAM et al. submitted). Actin filaments then become tightly bundled by plastin (human fimbrin), a 60-kD actin-binding protein (BRETSCHER 1981) which appears essential to cross-link and stabilize cytoskeletal projections that form the membrane bundles (ADAM et al. 1995).

We are still a long way from understanding the signalization pathways leading to *Shigella* entry into cells; however, some keystones have been recognized that should facilitate future progress. So far, the signals identified for *Shigella* appear different from those identified during *Salmonella* entry. Unlike what has been shown for *S. typhimurium* (GALAN et al. 1992b), there is no obvious tyrosine phosphorylation of the EGF receptor and no calcium flux upon entry of *Shigella* (CLERC et al. 1989). Finally, ruffle induction and bacterial entry appear to be rho independant for *S. typhimurium* (JONES et al. 1993), but not for *Shigella* (ADAM et al. submitted).

2.2 Genetics of Entry

Evidence for an essential role of a plasmid in entry came from the observation that a plasmid of about 200 kb was present in invasive isolates of *Shigella* and EIEC, and that deletions within or loss of this plasmid abolished entry (SANSONETTI et al. 1981, 1982, 1983a; HARRIS et al. 1982). Moreover, mobilization of the large plasmid from *S. flexneri* to *E. coli* K12 gave rise to a strain which was able to enter HeLa cells (SANSONETTI et al. 1983b). The Ipa (invasion plasmid antigen) proteins (A, 71 kD; B, 62 kD; C, 42 kD; D, 39 kD) were the first virulence plasmid-encoded proteins to be identified (HALE et al. 1985; OAKS et al. 1986; BUYSSE et al. 1987). Cloning into a cosmid and transposon mutagenesis were then used to identify entry genes (MAURELLI et al. 1985; SASAKAWA et al., 1986, 1988; KATO et al. 1989).

The nucleotide sequence of the 30.5-kb region that is necessary and apparently sufficient for entry of *Shigella* into epithelial cells has now been determined

(ADLER et al. 1989; ALLAOUI et al. 1992a,b, 1993a,b, 1995, and unpublished data; ANDREWS and MAURELLI 1992; BAUDRY et al. 1988; BUYSSE et al. 1990; SASAKAWA et al. 1989, 1993; VENKATESAN et al. 1988, 1992; VENKATESAN and BUYSSE 1991). In addition to the gene for a transcriptional activator (*virB*, see below), the entry region contains 33 genes clustered in two loci which are transcribed in opposite orientation (Fig. 2). One locus encodes secretory proteins, the Ipa proteins, and their molecular chaperone, IpgC, and the other locus encodes a specialized secretion apparatus, the Mxi-Spa translocon.

2.2.1 The Mxi-Spa Translocon

Secretion of IpaB and IpaC into the culture medium was demonstrated by ANDREWS et al. (1991), and analysis of concentrated culture supernatant fluids by SDS-PAGE and Coomassie blue staining indicated that wild-type *S. flexneri* secretes about ten polypeptides into the growth medium (ALLAOUI et al. 1992b). In addition to IpaB and IpaC, IpaA and IpaD have been detected in the culture medium (MÉNARD et al. 1993, 1994a,b). Characterization of the phenotype of non-invasive mutants obtained after transposon mutagenesis or constructed by allelic replacement led to the identification of a locus whose products are required for surface presentation and secretion of Ipa antigens (ANDREWS et al. 1991; ALLAOUI et al. 1992b, 1993a, 1995; VENKATESAN et al. 1992; SASAKAWA et al. 1993). This locus contains two adjacent operons clustering about 20 *mxi* (membrane excretion of *Ipa*) and *spa* (surface presentation of invasion plasmid antigens) genes (Fig. 2). Since Ipa proteins are necessary for entry (see below), the noninvasive phenotype of *mxi* and *spa* mutants is most likely a consequence of their secretion defect.

Homologues of more than ten Mxi and Spa proteins are required for secretion of Yop proteins by *Yersinia* (MICHIELS et al. 1991; ALLAOUI et al. 1994; BERGMAN et al. 1994; FIELDS et al. 1994; WOESTYN et al. 1994) and for cell invasion by *Salmonella* (GALAN et al. 1992a; GINOCCHIO et al. 1992; GROISMAN and OCHMAN 1993; EICHELBERG et al. 1994). In addition, similar proteins have been found in several plant pathogens, leading to the proposal that these related proteins are components of a Sec-independent secretion pathway, which was designated the type-III secretion pathway (reviewed by VAN GIJSEGEM et al. 1993).

Little is known about the localization of Mxi and Spa proteins. MxiJ and MxiM are lipoproteins and might be anchored to the outer membrane by their

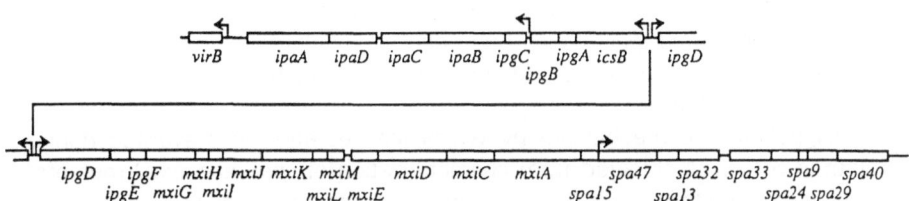

Fig. 2. Genetic organization of entry region

terminal lipid moiety (ALLAOUI et al. 1992b). MxiA has been detected in the inner membrane and might contain a N-terminal domain composed of six trans-membrane segments and a cytoplasmic C-terminal domain (ANDREWS et al. 1991; ANDREWS and MAURELLI 1992). The presence of hydrophobic regions in Spa9, Spa15, Spa24, Spa29, and Spa40 suggests that these proteins might also be located in the inner membrane (VENKATESAN et al. 1992; SASAKAWA et al. 1993). MxiD was tentatively localized in the outer membrane (ALLAOUI et al. 1993b), and MxiG was shown to be associated to both the inner and outer membranes (ALLAOUI et al. 1995).

2.2.2 The Ipa Invasins

The role of the *ipa* operon (Fig. 2) in entry was first investigated using trans-poson insertion mutants constructed on the virulence plasmid (MAURELLI et al. 1985; SASAKAWA et al. 1986; WATANABE et al. 1990). Then, each of the *ipaB*, *ipaC*, and *ipaD* genes carried by the virulence plasmid was inactivated by allelic replacement with a gene mutagenized in vitro by insertion of a non-polar cassette (HIGH et al. 1992; MÉNARD et al. 1993). The *ipaB*, *ipaC*, and *ipaD* mutants were unable to induce actin polymerization at the site of attachment of bacteria to the cellular membrane and were therefore unable to enter. Since *ipa* mutants are defective in entry but not in secretion of the remaining Ipa proteins, which differentiates them from *mxi* and *spa* mutants, IpaB, IpaC, and IpaD appear as potential effectors of *Shigella* entry into epithelial cells (MÉNARD et al. 1993). Moreover, when internalized by macrophages, *ipa* mutants remain trapped in the phagosome and are not cytotoxic (ZYCHLINSKY et al. 1994b). This suggests that Ipa proteins are also involved in lysis of the endosome membrane upon entry into epithelial cells.

A complex containing IpaB, IpaC, and probably IpaA has been detected in the extracellular milieu of *Shigella* (MÉNARD et al. 1994b). This complex is likely to play a role in entry, since the Ipa proteins are not present at the bacterial surface. Within the cytoplasm, IpaB and IpaC are each associated to a molecular chaper-one, IpgC, which is encoded by the gene located immmediately upstream from *ipaB* (Fig. 2) (MÉNARD et al. 1994b). IpgC is necessary to stabilize IpaB and to prevent its association to IpaC. The rationale for the requirement of partitioning IpaB and IpaC in the cytoplasm might be that production and secretion of Ipa proteins are uncoupled. Indeed, although the ability to secrete the Ipa proteins is essential for entry, only a small fraction (about 10%) of these proteins are actually secreted by *Shigella* growing in laboratory media. It has recently been shown that secretion of IpaB and IpaC which had been accumulated in the cytoplasm of in vitro grown bacteria was induced upon contact of *Shigella* with epithelial cells (MÉNARD et al. 1994a).

Insight into the mechanism by which Ipa secretion is prevented during growth in vitro was obtained from characterization of *ipaB* and *ipaD* mutants, which were found to release all the remaining Ipas into the medium (MÉNARD et al. 1994a). A small fraction of IpaB and IpaD is associated to the membrane, perhaps with (or within) the secretion apparatus. Moreover, *ipaB* and *ipaD* mutations also lead to enhanced secretion of a set of about 15 proteins, some of which are hardly detectable in the growth medium of the wild-type strain (PARSOT et al. 1995). Therefore, initial characterization of proteins secreted by *Shigella* during exponential growth in vitro, a condition which is now known to inhibit the activity of the Mxi-Spa secretion apparatus, has underestimated the number of *Shigella* secretory proteins. Although likely, induced secretion of these proteins upon contact of bacteria with epithelial cells and their involvement in pathogenicity remain to be ascertained.

2.2.3 Transcriptional Regulation

Expression of the *ipa*, *mxi*, and *spa* operons is controlled by a regulatory cascade involving two transcriptional activators, VirB and VirF, which are encoded by the large plasmid, and the product of a chromosomal gene, *virR*. Transposon insertions in *virB* (*ipaR*, *invE*), which is located immediately downstream from the *ipa* operon (Fig. 2), abolish transcription of the *ipa*, *mxi*, and *spa* operons (MAURELLI et al. 1985; SASAKAWA et al. 1988; ADLER et al. 1989; BUYSSE et al. 1990; WATANABE et al. 1990). Expression of *virB* is positively regulated by *virF*, which is located away from the entry region and encodes a transcriptional activator of the AraC family (SAKAI et al. 1986a,b 1988; ADLER et al. 1989; TOBE et al. 1991, 1993).

Entry of *Shigella* is regulated by the growth temperature; strains are invasive when grown at 37°C and noninvasive when grown at 30°C (MAURELLI et al. 1984). The *virR* gene was identified following Tn*10* mutagenesis of a *S. flexneri* strain carrying a transcriptional *mxi-lacZY* fusion and selection for a Lac+ phenotype at 30°C (MAURELLI and SANSONETTI 1988). Transduction of the *virR*: Tn*10* mutation to wild-type *Shigella* resulted in a strain that was invasive at both 30°C and 37°C, confirming that *virR* is involved in the temperature-regulated expression of the invasion phenotype. The *virR* mutation is allelic to *osmZ*, *drdX*, *bglY*, and *pilG* mutations identified in *E. coli* (DORMAN et al. 1990; GÖRANSSON et al. 1990; HULTON et al. 1990; MAY et al. 1990). The corresponding wild-type gene encodes the histone-like protein H1 (H-NS), which might be responsible for local changes in DNA supercoiling leading to extinction of gene expression under certain growth conditions.

In addition to temperature, osmolarity modulates invasion gene expression. In *E. coli*, *envZ* and *ompR* encode a two-component regulatory system that controls transcription of *ompF* and *ompC* in response to change in osmolarity of the growth medium (COMEAU et al. 1985). Expression of an *mxi-lac* transcriptional fusion was enhanced in high-osmolarity conditions and reduced in *envZ* and *Δ(ompR-envZ)* mutants. Transduction of these mutations to wild-type *Shigella* gave rise to mutants that were less invasive than the wild-type strain (BERNARDINI et al. 1990).

3 Intra- and Intercellular Dissemination

3.1 Biology of Intracellular Movement

While extracellular *Shigella* are nonmotile, intracellular bacteria move to occupy the entire cytoplasm of infected cells and to spread from cell to cell. Two apparently independent movements have been described in different cell lines.

3.1.1 The Ics Movement

The initial observation that *Shigella* move within the cytoplasm of infected cells came from phase-contrast microcinematography studies; movement of bacteria was random and sometimes led to formation of structures protruding from the cell surface and containing bacteria at their tip (OGAWA et al. 1968). Ultrastructural analysis of these protrusions indicated that they have a diameter of about 0.5 µm and a length of up to 20 µm (KADURUGAMUWA et al. 1991; PREVOST et al. 1992; SANSONETTI et al. 1994). Bacteria in these protrusions are located at the top of tightly packed actin filaments which, in some instances, appear to form a cylinder. Pictures showing a protrusion extending from one cell and penetrating into the adjacent cell indicated that these protrusions can allow passage of *Shigella* from cell to cell without release of bacteria into the extracellular medium. Genetic studies (see below) confirmed the importance of this movement, designated Ics (*intra-* and *inter*cellular spread), in the dissemination of bacteria from the primary infected cell to adjacent cells.

The Ics movement is inhibited by treatment with cytochalasin D that prevents polymerization of monomeric actin (G-actin) into filaments (PAL et al. 1989; BERNARDINI et al. 1989). In addition to actin, several cellular proteins such as vinculin and plastin, but not myosin, are associated with the polymerized structure that trails behind the intracellular bacteria (KADURUGAMUWA et al. 1991; PREVOST et al. 1992). Using a cell line which does not produce cell adhesion molecules (CAM) and transfectants expressing either L-CAM or N-cadherin, it was shown that CAMs are required for cell-to-cell spread of *Shigella* (SANSONETTI et al. 1994). Cadherin was important for both the structural organization of the protrusion and the internalization of a protrusion by an adjacent cell. Moreover, L-CAM, α actinin, vinculin, and α- and β-catenins are associated with protrusions that initiate at intermediate junctions.

3.1.2 The Olm Movement

In chicken embryo fibroblasts, which have a highly organized cytoskeleton, intracellular bacteria interact with and progress along stress fibers, a movement that was designated Olm (organelle-like movement) (VASSELON et al. 1991). Movement of bacteria along the actin filament ring of the perijunctional area has also been observed in infected Caco-2 cells (VASSELON et al. 1992).

3.2 Genetics of Intercellular Dissemination

Dissemination of bacteria from cell to cell without release into the extracellular medium is reflected by formation of plaques on a confluent cell monolayer (OAKS et al. 1985), a phenotype which has been very useful in the identification of bacterial genes involved in this process.

3.2.1 Intracellular Movement

Transposon insertions in *virG*, which is located away from the entry region on the virulence plasmid pMYSH6000 of *S. flexneri* 2a, do not affect entry but abolish cell-to-cell spread (MAKINO et al. 1986; LETT et al. 1989). A similar gene, designated *icsA*, has been characterized on the virulence plasmid pWR100 of *S. flexneri* 5; the *icsA* mutant is unable to elicit actin polymerization at the poles of bacteria and to induce formation of protrusions (BERNARDINI et al. 1989). The *icsA* (*virG*) gene, whose expression is regulated by VirF (SAKAI et al. 1988; ADLER et al. 1989), encodes a 130-kD outer membrane protein (LETT et al. 1989; BERNARDINI et al. 1989; D'HAUTEVILLE and SANSONETTI 1992). The purified protein bound and hydrolyzed ATP, which suggests that ATP hydrolysis might be involved in bacterial movement (GOLDBERG et al. 1993). In bacteria grown in vitro, IcsA is located at the distal poles of dividing bacteria, a distribution which was also observed in intracellular bacteria after infection of HeLa cells (GOLDBERG et al. 1993). This unipolar distribution of IcsA correlates with the unipolar reorganization of F-actin seen on the surface of dividing bacteria at the onset of the Ics movement (PREVOST et al. 1992).

About 50% of IcsA is released into the culture medium as a 95-kD species (GOLDBERG et al. 1993). The secreted protein is cleaved after residue Ala-52. The difference in size between the 95-kD secreted protein and the 130-kD gene product suggested that a second cleavage occurs in the C-terminal portion of the precursor, and a polypeptide of 37-kD corresponding to the C-terminal part of IcsA was indeed detected in whole-cell extracts (NAKATA et al. 1992, 1993). Cleavage of the C-terminal portion of the precursor prior to release of the mature protein into the culture medium is reminiscent of the secretion mechanism of *Neisseria gonorrhoeae* IgA1 protease (POHLNER et al. 1987). Mutations in *virK*, which is also located on the virulence plasmid, decrease the amount of the cell-associated 130-kD IcsA protein, but not that of the 37-kD C-terminal fragment (NAKATA et al. 1992).

3.2.2 Cell-to-Cell Spread

Due to their inability to induce actin polymerization and the formation of protrusions, *icsA* (and *virK*) mutants are defective in intracellular movement and therefore in intercellular dissemination. Recently acquired data point to genes whose inactivation does not abolish the intracellular movement but affects subsequent steps in cell-to-cell spread.

The *mxiG* gene (Fig. 2) is required for Ipa secretion and therefore for entry. Complementation of an *mxiG* null mutant by a plasmid carrying a wild-type copy of *mxiG* restored Ipa secretion, entry into HeLa cells, and intercellular dissemination. The complementing plasmid has been subjected to site-directed mutagenesis which led to the characterization of an *mxiG* allele (*mxiG**) that complemented the *mxiG* null mutant for entry but not for cell-to-cell spread (ALLAOUI et al. 1995). This suggests that MxiG and possibly proteins secreted by the Mxi-Spa translocon are required not only for entry but also for intercellular dissemination.

Once a protrusion extending from an infected cell is engulfed by an adjacent cell, bacteria within this protrusion are surrounded by two cellular membranes, that of the protrusion itself and that of the cell into which the protrusion enters (Fig. 1). Lysis of these two membranes allows bacteria to gain access to the cytoplasm of the newly infected cell, thereby completing the process of intercellular dissemination. Inactivation of *icsB*, which is located in the entry region (Fig. 2), gave rise to a strain that did not spread from cell to cell although it was able to induce the formation of protrusions (ALLAOUI et al. 1992a). The *icsB* mutant remained trapped in vacuoles surrounded by the two membranes. The fact that the *icsB* mutant was able to lyse the endosome membrane during entry indicates that different membranolytic activities are required for escape from the endosome and from the protrusion.

4 Tissue Invasion

4.1 Portal of Entry

S. flexneri enters polarized cells through their basolateral pole rather than their apical pole (MOUNIER et al. 1992). This observation raises the question of the access of *Shigella* to the basolateral pole of enterocytes in the mucosa. The definition of the site of entry of *Shigella* is of importance, particularly considering that mucosal destruction and presumably invasion occur only in the colon. Using the rabbit intestinal loop model, WASSEF et al. (1989) showed that both invasive and noninvasive *Shigella* were phagocytosed by M cells over lymphoid follicles of Peyer's patches. The invasive strain appeared to escape from the phagocytic vacuole and replicate intracellularly. The M cell may thus serve as a preferential site of entry for *Shigella*, from which bacteria could either disseminate to adjacent enterocytes by expressing the Ics phenotype or, following transcytosis through or lysis of M cells, invade enterocytes by their basolateral pole. This concept is supported by the observation that small nodular abscesses induced by the *icsA* mutant in macaque monkeys were located over lymphoid follicles (SANSONETTI et al. 1991). Since the *icsA* is unable to spread from cell to cell, sites of tiny ulcerations should correspond to sites of entry into the epithelium.

4.2 Invasion and Inflamation

Recent data indicate that intestinal inflammation, in addition to causing destruction of intestinal tissues as a consequence of *Shigella* invasion, may also account for epithelial and mucosal invasion. Infection of resident macrophages present in the dome of the lymphoid structures associated with the mucosa may be responsible for early inflammation. In vitro, macrophages pretreated by LPS, thus mimicking a situation which is likely to prevail in follicular domes, and subsequently infected with invasive *S. flexneri* initiate a cell death program (ZYCHLINSKY et al. 1992) and, before apoptosis occurs, release massive amounts of mature IL-1β (ZYCHLINSKY et al. 1994a). This may represent an early signal causing a local cascade of release of cytokines (i.e., TNFα) and chemokines (i.e., IL-8) which recruit polymorphonuclear (PMN) leukocytes in the follicular and parafollicular area. The possible role of early inflammation in destabilizing the intestinal epithelium and facilitating enterocyte invasion has been confirmed in vitro (PERDOMO et al. 1994b). Apical *S. flexneri* was shown to attract and induce migration of basal PMN through a confluent monolayer of human colonic T-84 cells by a CR3-dependent process. This transmigration appeared to be necessary for bacteria to enter into the monolayer and invade T-84 cells through their basolateral pole. Two recent series of experiments carried out in vivo in the rabbit ligated loop assay confirm this model. Prior to infection, treatment of animals with an anti-CD18 monoclonal antibody that neutralizes immigration of PMN leukocytes into tissues not only controlled inflammation, but also decreased epithelial invasion (PERDOMO et al. 1994a). In addition, treatment of animals with IL-1ra (AREND 1993) during infection also significantly decreased inflammation and tissue invasion (SANSONETTI et al. submitted).

Shigellosis appears to be similar in its pathogenesis to certain forms of acute inflammatory bowel diseases (MATHAN and MATHAN 1986), particularly ulcerative colitis, where IL-1 seems to play a critical role (SARTOR 1991). Recent immunohistochemical analysis of rectal biopsies taken from patients with shigellosis has shown, in tissue samples characterized by severe inflammation, a predominance of such IL-1-producing cells as mononuclear cells, PMN leukocytes, and endothelial cells (RAQUIB et al. 1995). Invasion of lymphoid follicles is likely to be the trigger of this inflammatory process, which then subverts the epithelial architecture, thereby causing extensive invasion of the intestine.

5 Conclusion

Invasion of a cell monolayer by *Shigella* involves numerous interactions between the bacterium and the host cell (Fig. 1). A systematic genetic analysis of the virulence plasmid, together with the localization of most virulence factors which have been identified so far, suggests the following model for the cascade of events which leads to the delivery of *Shigella* invasins onto the epithelial cell surface and to dissemination of intracellular bacteria from cell to cell.

In bacteria growing in vitro at 37°C, binding of VirF on the *virB* promoter induces *virB* expression, which allows the VirB protein to activate the *ipa*, *mxi*, and *spa* promoters. Expression of the *mxi* and *spa* operons leads to assembly of the Mxi-Spa translocon, albeit in an inactive form. Under these conditions, expression of the *ipa* operon leads to accumulation of Ipa proteins in the cytoplasm, where IpaB and IpaC are each associated to a molecular chaperone, IpgC, which prevents their premature association. In addition, a small fraction of IpaB and IpaD is present in the membrane and acts as a cork for the translocon. Contact of *Shigella* with the cell surface activates the translocon, which results in secretion of presynthesized Ipa proteins. IpaB, IpaC, and IpaA associate in the extracellular milieu, and the Ipa complex would interact with the cellular membrane to induce a cascade of cellular signalization which ultimately leads to internalization of the bacterium. The Ipa proteins appear to also be required for escape from the endosome. Bacterial multiplication within the cellular cytoplasm allows the unipolar distribution of surface-bound IcsA, which recruits some actin nucleator(s) or bundling protein(s) and allows the formation of protrusions. Cellular adhesion molecules are required not only for proper structure of protrusions but also for internalization of protrusions by adjacent cells, which suggests that protrusions might be actively endocytosed during intercellular spread. The nature of bacterial components that could be involved in this other type of induced phagocytosis remains to be determined. Although characterization of *mxi*, *spa*, and *ipa* genes has hitherto focused on their role in entry, some of these genes might be also involved in intercellular dissemination.

However, this "phenotypic cassette" needs to be considered in the perspective of the in vivo situation, in which all components of the infectious process including nonspecific immune defenses of the host are present. This approach requires permanent movement between cell assay systems and animal models of infection, as well as attempts at reconstituting limited compartments of the intestinal barrier by mixing in vitro some of its key-cell constituents, thus allowing the isolation and study of specific signals leading to subversion of the structure. So far, three major points emerge from the combination of these studies: the key role of M cells—present in the follicular-associated dome of rectal and colonic mucosal lymphoid structures-in allowing crossing of the epithelial barrier by the pathogen; the importance of the inflammatory reaction elicited early in these structures and beyond in disturbing the integrity of the epithelial barrier, thus increasing access of the bacteria to epithelial cells; the essential role of the invasive phenotype that has been described in vitro which allows efficient invasion and colonization of the epithelial lining. Synergy between invasion, intracellular colonization, and inflammation is likely to account for a disease which appears to combine the classical elements of an infection with striking aspects of inflammatory bowel diseases (IBD) such as ulcerative colitis. These elements are summarized in Fig. 3. How much the understanding of the pathogenesis of shigellosis will profit from studies on IBD and vice versa will be seen in the near future.

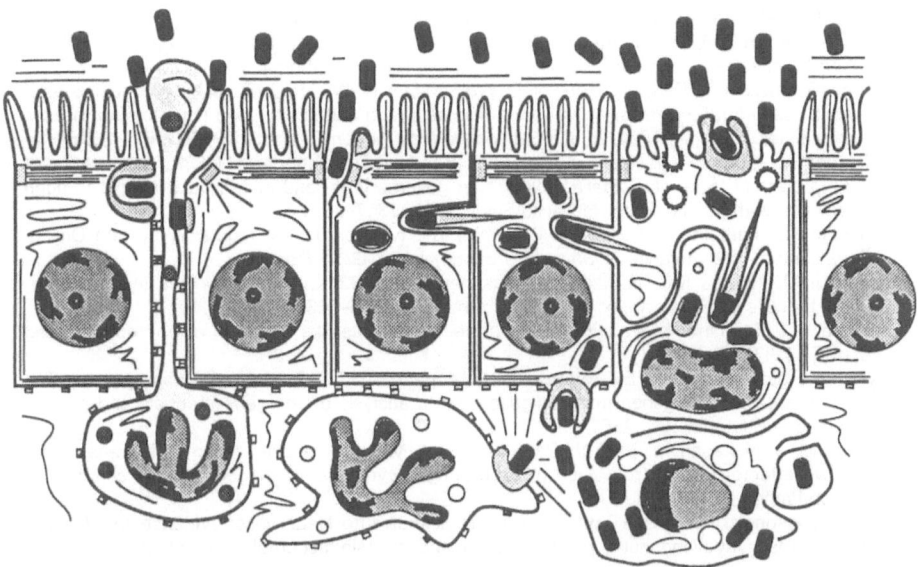

Fig. 3. Current simplified view of the pathogenesis of shigellosis. Entry of the pathogen via M cells, apoptotic killing of macrophages, elicitation of inflammatory reaction, transmigration of in situ-attracted polymorphonuclear leukocytes through the epithelial lining, and facilitation of bacterial access to basolateral pole of epithelial cells, entry, and cell-to-cell spread of bacteria

References

Adam T, Arpin M, Prévost MC, Gounon P, Sansonetti PJ (1995) Cytoskeletal rearrangements and functional role of T-plastin during entry of *Shigella flexneri* into HeLa cells. J Cell Biol (in press)

Adler B, Sasakawa C, Tobe T, Makino S, Komatsu K, Yoshikawa M (1989) A dual transcriptional activation system for the 230 kb plasmid genes coding for virulence-associated antigens of *Shigella flexneri*. Mol Microbiol 3: 627–635

Allaoui A, Mounier J, Prévost MC, Sansonetti PJ, Parsot C (1992a) *icsB*: a *Shigella flexneri* virulence gene necessary for the lysis of protrusions during intercellular spread. Mol Microbiol 6: 1605–1616

Allaoui A, Sansonetti PJ, Parsot C (1992b) MxiJ, a lipoprotein involved in secretion of *Shigella* Ipa invasins, is homologous to YscJ, a secretion factor of the *Yersinia* Yop proteins. J Bacteriol 174: 7661–7669

Allaoui A, Ménard R, Sansonetti PJ, Parsot C (1993a) Characterization of the *Shigella flexneri ipgD* and *ipgF* genes, which are located in the proximal part of the *mxi* locus. Infect Immun 61: 1717–1714

Allaoui A, Sansonetti PJ, Parsot C (1993b) MxiD: an outer membrane protein necessary for the secretion of the *Shigella flexneri* Ipa invasins. Mol Microbiol 7: 59–68

Allaoui A, Woestyn S, Sluiters C, Cornelis GR (1994) YscU, a *Yersinia entercolitica* inner membrane protein involved in Yop secretion. J Bacteriol 176: 4534–4542

Allaoui A, Sansonetti PJ, Ménard R, Barzu S, Mounier J, Phalipon A, Parsot C (1995) MxiG, a membrane protein required for secretion of *Shigella* invasins: involvement in entry into epithelial cells and in intracellular dissemination. Mol Microbiol (in press)

Andrews GP, Hromockyj AE, Coker C, Maurelli AT (1991) Two novel virulence loci, *mxiA* and *mxiB*, in *Shigella flexneri* 2a facilitate excretion of invasion plasmid antigens. Infect Immun 59: 1997–2005

Andrews GP, Maurelli AT (1992) *mxiA* of *Shigella flexneri* 2a, which facilitates export of invasion plasmid antigens, encodes a homolog of the low-calcium response protein, LcrD, or *Yersinia pestis*. Infect Immun 60: 3287–3295

Arend WP (1993) Interleukin-1 receptor antagonist. Adv Immunol 54: 167–227

Baudry B, Kaczorek M, Sansonetti PJ (1988) Nucleotide sequence of the invasion plasmid antigen B and C genes (ipaB and ipaC) of Shigella flexneri. Microb Pathog 4: 345–357

Bergman T, Erickson K, Galyov E, Persson C, Wolf-Watz H (1994) The lcrB (yscN/U) gene cluster of Yersinia pseudotuberculosis is involved in Yop secretion and shows high homology to the spa gene cluster of Shigella flexneri and Salmonella typhimurium. J Bacteriol 176: 2619–2626

Bernardini ML, Mounier J, d'Hauteville H, Coquis-Rondon M, Sansonetti PJ (1989) Identification of icsA, a plasmid locus of Shigella flexneri that governs intra- and inter-cellular spread through interaction with F-actin. Proc Natl Acad Sci USA 86: 3867–3871

Bernardini ML, Fontaine A, Sansonetti PJ (1990) The two-component regulatory system OmpR-EnvZ controls the virulence of Shigella. J Bacteriol 172: 6274–6281

Bernardini ML, Sanna MG, Fontaine A, Sansonetti PJ (1993) OmpC is involved in invasion of epithelial cells by Shigella flexneri. Infect Immun 61: 3625–3635

Bretscher A (1981) Fimbrin is a cytoskeletal protein that crosslinks F-actin in vitro. Proc Natl Acad Sci USA 78: 6849–6853

Buysse JM, Stover CK, Oaks EV, Venkatesan M, Kopecko DJ (1987) Molecular cloning of invasion plasmid antigen (ipa) genes from Shigella flexneri analysis of ipa gene products and genetic mapping. J Bacteriol 169: 2561–2569

Buysse JM, Venkatesan M, Mills J, Oaks EV (1990) Molular characterization of a trans-acting, positive effector (ipaR) of invasion plasmid antigen synthesis in Shigella flexneri serotype 5. Microb Pathog 8: 197–211

Clerc P, Sansonetti PJ (1987) Entry of Shigella flexneri into HeLa cells: evidence for directed phagocytosis involving actin polymerization and myosin accumulation. Infect Immun 55: 2681–2688

Clerc P, Berthon B, Claret M, Sansonetti PJ (1989) Internalization of Shigella flexneri into HeLa cells occurs without an increase in cytosolic Ca^{2+} concentration. Infect Immun 57: 2919–2922

Comeau DE, Ikenaka K, Tsung K, Inouye M (1985) Primary characterization of the protein products of the Escherichia coli ompB locus: structure and regulation of synthesis of the OmpR and EnvZ proteins. J Bacteriol 164: 578–584

Dehio C, Prévost MC, Sansonetti PJ (1995) Invasion of epithelial cells by Shigella flexneri induces tyrosine phosphorylation of cortactin by a $pp60^{c-src}$ signalling pathway. EMBO J (in press)

d'Hauteville H, Sansonetti PJ (1992) Phosphorylation of IcsA by cAMP-dependent protein kinase and its effect on intercellular spread of Shigella flexneri. Mol Microbiol 6: 833–841

Dorman CJ, Ni Bhriain N, Higgins CF (1990) DNA supercoiling and environmental regulation of virulence gene expression in Shigella flexneri. Nature 344: 789–792

DuPont HL, Formal SB, Hornick RB, Snyder MJ, Libonati JB, Sheahan DG, LaBrec EH, Kalas JP (1971) Pathogenesis of Escherichia coli diarrhea. N Engl J Med 285: 1–9

DuPont HL, Levine MM, Hornick RB, Formal SB (1989) Inoculum size in shigellosis and implications for expected mode of transmission. J Infect Dis 159: 1126–1128

Fields KA, Plano GV, Straley SC (1994) A low-Ca^{2+} response (LCR) secretion (ysc) locus lies within the lcrB region of the LCR plasmid in Yersinia pestis. J Bacteriol 176: 569–579

Francis CL, Ryan TA, Jones BD, Smith SJ, Falkow S (1993) Ruffles induced by Salmonella and other stimuli direct macropinocytosis of bacteria. Nature 364: 639–642

Galan JE, Ginocchio C, Costeas P (1992a) Molular and functional characterization of the Salmonella invasion gene invA: homology of InvA to members of a new protein family. J Bacteriol 174: 4338–4349

Galan JE, Pace J, Hayman MJ (1992b) Involvement of the epidermal growth factor receptor in the invasion of cultured mammalian cells by Salmonella typhimurium. Nature 357: 588–589

Ginocchio CC, Pace J, Galan JE (1992) Identification and molecular caracterisation of a Salmonella typhimurium gene involved in triggering the internalization of Salmonella into cultured epithelial cells. Proc Natl Acad sci USA 89: 5976–5980

Gerber DF, Watkins HMS (1961) Growth of shigellae in monolayer tissue cultures. J Bacteriol 82: 815–822

Goldberg MB, Bârzu O, Parsot C, Sansonetti PJ (1993) Unipolar localization and ATPase activity of IcsA, a Shigella flexneri protein involved in intracellular movement. J Bacteriol 175: 2189–2196

Göransson M, Sondén B, Nilsson P, Dagberg B, Forsman K, Emanuelson K, Uhlin BE (1990) Transcriptional silencing and thermoregulation of gene expression in Escherichia coli. Nature 344: 682–685

Gots R, Formal SB, Giannella RA (1974) Indomethacin inhibition of Samonella typhimurium, Shigella flexneri, and cholera-mediated rabbit ileal secretion. J Infect Dis 130: 280–284

Groisman EA, Ochman H (1993) Cognate gene clusters govern invasion of host epithelial cells by *Salmonella typhimurium* and *Shigella flexneri*. EMBO J 12: 3779–3787

Hale TL (1991) Genetic basis of virulence in *Shigella* species. Microbiol Rev 55: 206–224

Hale TL, Morris RE, Bonventre PF (1979) *Shigella* infection of Henle intestinal epithelial cells: role of the host cell. Infect Immun 24: 887–894

Hale TL, Oaks, EV, Formal SB (1985) Identification and antigenic characterization of virulence-associated, plasmid-coded proteins of *Shigella* spp. and enteroinvasive *Escherichia coli*. Infect Immun 50: 620–629

Harris JR, Wachmuth IK, Davies BR, Cohen ML, (1982) High molecular weight plasmid correlates with *Escherichia coli* enteroinvasiveness. Infect Immun 37: 1295–1298

High N, Mounier J, Prévost MC, Sansonetti PJ (1992) IpaB of *Shigella flexneri* causes entry into epithelial cells and escape from the phagocytic vacuole. EMBO J 11: 1991–1999

Hulton CSJ, Seirafi A, Hinton JCD, Sidebotham JM, Waddell L, Pavitt GD, Owen-Hughes T, Spassky A, Buc H, Higgins CF (1990) Histone-like protein H1 (H-NS), DNA supercoiling, and gene expression in bacteria. Cell 63: 631–642

Isberg RR, Leong JM (1990) Multiple β1 chain integrins are receptors for invasin, a protein that promotes bacterial penetration into mammalian cells. Cell 60: 861–871

Jones BD, Paterson HF, Hall A, Falkow S (1993) *Salmonella typhimurium* induces membrane ruffling by a growth factor-receptor-independent mechanism. Proc Natl Acad Sci USA 90: 10390–10394

Kadurugamuwa JL, Rhode M, Wehland J, Timmis KN (1991) Intercellular spread of *Shigella flexneri* through a monolayer mediated by membranous protrusions and associated with reorganization of the cytoskeletal protein vinculin. Infect Immun 59: 3463–3471

Kato J, Ito K, Nakamura A, Watanabe H (1989) Cloning of regions required for contact hemolysis and entry into LLC-MK2 cells from *Shigella sonnei* form I plasmid: *virF* is a positive regulator gene for these phenotypes. Infect Immun 57: 1391–1398

LaBrec EH, Schneider H, Magnani TJ, Formal SB (1964) Epithelial cell penetration as an essential step in the pathogenesis of bacillary dysentery. J Bacteriol 88: 1503–1518

Lett MC, Sasakawa C, Okada N, Sakai T, Makino S, Yamada M, Komatsu K, Yoshikawa M (1989) *virG*, a plasmid-coded virulence gene of *Shigella flexneri*: identification of the *virG* protein and determination of the complete coding sequence. J Bacteriol 171: 353–359

Makino S, Sasakawa C, Kamata K, Kurata T, Yoshikwa M (1986) A virulence determinant required for continuous reinfection of adjacent cells on large plasmid in *Shigella. flexneri* 2a. Cell 46: 551–555

Mathan MM, Mathan VI (1986) Ultrastructural pathology of the rectal mucosa in *Shigella* dysentery. Am J Pathol 123: 25–38

Maurelli AT, Baudry B, d'Hauteville H, Hale TL, Sansonetti PJ (1985) Cloning of plasmid DNA sequences involved in invasion of HeLa cells by *Shigella Flexneri*. Infect Immun 49: 164–171

Maurelli AT, Blackmon B, Curtiss R (1984) Temperature-dependent expression of virulence genes in *Shigella* species. Infect Immun 43: 195–201

Maurelli AT, Sansonetti PJ (1988) Identification of a chromosomal gene controlling temperature-regulated expression of *Shigella* virulence. Proc Natl Acad Sci USA 85: 2820–2824

May G, Dersch P, Haardt M, Middendorf A, bremer E (1990) The *osmZ* (*bglY*) gene encodes the DNA-binding protein H-NS (H1a), a component of the *Escherichia coli* K12 nucleoid. Mol Gen Genet 224: 81–90

Ménard R, Sansonetti PJ, Parsot C (1993) Nonpolar mutagenesis of the *ipa* genes defines IpaB, IpaC, and IpaD as effectors of *Shigella flexneri* entry into epithelial cells. J Bacteriol 175: 5899–5906

Ménard R, Sansonetti PJ, Parsot C (1994a) The secretion of the *Shigella Flexneri* Ipa invasins is induced by the epithelial cell and controlled by IpaB and IpaD. EMBO J 13: 5293–5302

Ménard R, Sansonetti PJ, Parsot C, Vasselon T (1994b) Extracellular association and cytoplasmic partioning of the IpaB and IpaC invasins of *Shigella flexneri*. Cell 79: 515–525

Michiels T, Vanooteghem JC, Lambert de Rouvroy C, China B, Gustin, A, Boudry P, Cornelis GR (1991) Analysis of *virC*, an operon involved in secretion of Yop protein by *Yersinia enterocolitica*. J Bacteriol 173: 4994–5009

Mounier J, Vasselon T, Hellio R, Lesourd M, Sansonetti P (1992) *Shigella flexneri* enters human colonic Caco-2 epithelial cells through the basolateral pole. Infect Immun 60: 237–248

Nakata N, Sasakawa C, Okada N, Tobe T, Fukuda I, Suzuki T, Komatsu K, Yoshikawa M (1992) Identification and characterization of *virK*, a virulence-associated large plasmid gene essential for intercellular spreading of *Shigella flexneri*. Mol Microbiol 6: 2387–2395

Nakata N, Tobe T, Fukuda I, Suzuki T, Komatsu K, Yoshikawa M, Sasakawa C (1993) The absence of a surface protease, OmpT, determines the intercellular spreading ability of *Shigella*: the relationship between *ompT* and *kcpA* lock. Mol Microbiol 9: 459–468

Oaks EV, Wingfield ME, Formal SB (1985) Plaque formation by virulent *Shigella flexneri*. Infect Immun 48: 124–129

Oaks EV, Hale TL, Formal SB (1986) Serum immune response to *Shigella* protein antigens in rhesus monkeys and humans infected with *Shigella* spp. Infect Immun 53: 57–63

Ogawa H, Nakamura A, Nakaya R (1968) Cinemicrographic study of tissue culture infected with *Shigella flexneri*. Jpn J Med Sci Biol 21: 259–273

Okada N, Sasakawe C, Tobe T, Talukder KA, Komatsu K, Yoshika M (1991a) Construction of a physical map of the chromosome of *Shigella flexneri* 2a and the direct assignment of nine virulence-associated loci identified by Tn5 insertions. Mol Microbiol 5: 2171–2180

Okada N, Sasakawa C, Tobe T, Yamada M, Nagai S, Talukder KA, Komatsu K, Kanegasaki S, Yoshikawa M (1991b) Virulence-associated chromosomal loci of *Shigella flexneri* identified by random Tn5 insertion mutagenesis. Mol Microbiol 5: 187–195

Okamura N, Nagai T, Nakaya R, Kondo S, Murakami M, Hisatsune K (1983) HeLa cell invasiveness and O antigen of *Shigella flexneri* as separte and prerequisite attributes of virulence to evoke keratoconjuctivities in guinea pigs. Infect Immun 39: 505–513

Pal T, Newland JW, Tall BD, Formal SB, Hale TL (1989) Intracellular spread of *Shigella Flexneri* associated with the *kcpA* locus and a 140-kilodalton protein. Infect Immun 57: 477–486

Parsot C (1994) *Shigella flexneri*: genetics of entry and intercellular dissemination in epithelial cells. Curr Top Microbiol Immunol 192: 217–241

Parsot C, Ménard R, Gounon P, Sansonetti PJ (1995) Enhanced secretion through the *Shigella flexneri* Mxi-Spa translocon leads to assembly of extracellular protein into macromolecular structures. Mol Microbiol 16: 291–300

Perdomo JJ, Cavaillon JM, Huerre M, Ohayon H, Gounon P, Sansonetti PJ (1994a) Acute inflammation causes epithelial invaison and mucosal destruciton in experimental shigellosis. J Exp Med 180: 1307–1319

Perdomo JJ, Gounon P, Sansonetti PJ (1994b) Polymorphonuclear leukocyte (PMN) transmigration promotes invasion of colonic epithelial monolayer by *Shigella flexneri*. J Clin Invest 93: 633–643

Pohlner J, Halter R, Beyreuther K, Meyer TF (1987) Gene structure and extracellular secretion of *Neisseria gonorrhoeae* IgA protease. Nature 325: 458–462

Prévost MC, Lesourd M, Arpin M, Vernel F, Mounier J, Hellio R, Sansonetti PJ (1992) Unipolar reorganization of F-actin layer at bacterial division and bundling of actin filaments by plastin correlate with movement of *Shigella flexneri* within HeLa cells. Infect Immun 60: 4088–4099

Raquib R, Lindberg AA, Wretlind B, Bardhan PK, Andersson U, Andersson J (1995) Persistence of local cytokine production in shigellosis in acute and convalescent stages. Infect Immun 63: 289–296

Sakai T, Sasakawa C, Makino S, Kamata K, Yoshikawa M (1986a) Molecular cloning of a genetic determinant for Congo red binding ability which is essential for the virulence of *Shigella flexneri*. Infect Immun 51: 476–482

Sakai T, Sasakawa C, Makino S, Yoshikawa M (1986b) DNA sequence and product analysis of the *virF* locus responsible for Congo red binding and cell invasion in *Shigella flexneri* 2a. Infect Immun 54: 395–402

Sakai T, Sasakawa C, Yoshikawa M (1988) Expression of four virulence antigens of *Shigella flexneri* is positively regulated at the transcriptional level by the 30 kiloDalton VirF protein. Mol Microbiol 2: 589–597

Sansonetti PJ (1991) Genetic and Molecular basis of epithelial cell invaison by *Shigella* species. Rev Infect Dis 13: S285–S292

Sansonetti PJ (ed) (1992) Pathogenesis of shigellosis. Curr Top Microbiol Immunol 180

Sansonetti PJ, Mounier J (1987) Metabolic events mediating early killing of host cells infected by *Shigella flexneri*. Microb Pathog 3: 53–61

Sansonetti PJ, Kopecko DJ, Formal SB (1981) *Shigella sonnei* plasmids: evidence that a large plasmid is necessary for virulence. Infect Immun 34: 75–83

Sansonetti PJ, Kopecko DJ, Formal SB (1982) Involvement of a plasmid in the invasive ability of *Shigella flexneri*. Infect Immun 35: 852–860

Sansonetti PJ, d'Hauteville H, Ecobichon C, Pourcel C (1983a) Molular comparison of virulence plasmids in *Shigella* and enteroinvasive *Escherichia coli*. Ann Inst Pasteur Microbiol 134A: 295–318

Sansonetti PJ, Hale TL, Dammin GJ, Kapfer C, Collins HH Jr, Formal SB (1983b) Alterations in the pathogenicity of *Escherichia coli* K-12 after transfer of plasmid and chromosomal genes from *Shigella flexneri*. Infect Immun 39: 1392–1402

Sansonetti PJ, Ryter A, Clerc P, Maurelli AT, Mounier J (1986) Multiplication of *Shigella flexneri* within HeLa cells: lysis of the phagocytic vacuole and plasmid-mediated contact hemolysis. Infect Immun 51: 461–469

Sansonetti PJ, Arondel J, Fontaine A, d'Hauteville H, Bernardini ML (1991) *ompB* (osmo-regulation) and *icsA* (cell-to-cell spread) mutants of *Shigella flexneri*: vaccine candidates and probes to study the pathogenesis of shigellosis. Vaccine 9: 416–422

Sansonetti PJ, Mounier J, Prévost MC, Mege RM (1994) Cadherin expression is required for formation and internalization of *Shigella flexneri*-induced intercellular protrusions tht are involved in spread between epithelial cells. Cell 76: 829–839

Sartor RB (1991) Pathogenic and clinical relevance of cytokines in inflammatory bowel disease. Immunol Res 10: 465–471

Sasakawa C, Makino S, Kamata K, Yoshikawa M (1986) Isolation, characterization, and mapping of Tn5 insertions into the 140-megadalton invasion plasmid defective in the mouse Sereny test in *Shigella flexneri* 2a. Infect Immun 54: 32–36

Sasakawa C, Kamata K, Sakai T, Makino S, Yamada M, Okada N, Yoshikawa M (1988) Virulence-associated genetic regions comprising 31 kilobases of the 230-kilobase plasmid in *Shigella flexneri* 2a. J Bacteriol 170: 2480–2484

Sasakawa C, Adler B, Tobe T, Okada N, Naga S, Komatsu K, Yoshikawa M (1989) Functional organization and nucleotide sequence of the virulence region-2 on the large virulence plasmid in *Shigella flexneri* 2a. Mol Microbiol 3: 1191-1201

Sasakawa C, Komatsu K, Tobe T, Suzuki T, Yoshikawa M (1993) Eight genes in region 5 that form an operon are essential for invaison of epithelial cells by *Shigella flexneri* 2a. Infect Immun 175: 2334–2346

Sereny B (1957) Experimental keratoconjuctivis shigellosa. Acta Microbiol Acad Sci Hung 4: 367–376

Takeuchi A, Formal SB, Sprinz H (1968) Experimental acute colitis in the rhesus monkey following peroral infection with *Shigella flexneri*. Am J Pathol 52: 503–520

Tobe T, Nagai S, Okada N, Adler B, Yoshikawa M, Sasakawa C (1991) Temperature-regulated expression of invasion genes in *Shigella flexneri* in controlled through the transcriptional activation of the *virB* gene on the large plasmid. Mol Microbiol 5: 887–893

Tobe T, Yoshikawa M, Mizuno T, Sasakawa C (1993) Transcriptional control of the invasion regulatory gene *virB* of *Shigella flexneri*: activation by VirF and repression by H-NS. J Bacteriol 175: 6142–6149

Van Gijsegem F, Genin S, Boucher C (1993) Conservation of secretion pathways for pathogenicity determinants of plant and animal bacteria. Trends Microbiol. 1: 175–180

Vasselon T, Mounier J, Hellio R, Sansonetti PJ (1992) Movement along actin filaments of the perijunctional area and de novo polymerization of cellular actin are required for *Shigella flexneri* colonization of epithelial Caco-2 cell monolayers. Infect Immun 60: 1031–1040

Vasselon T, Mounier J, Prevost MC, Hellio R, Sansonetti PJ (1991) Stress fiber-based movement of *Shigella flexneri* within cells. Infect Immun 59: 1723–1732

Venkatesan MM, Buysse JM (1991) Nucleotide sequence of invasion plsmid antigen gene *ipaA* from *Shigella flexneri* 5. Nucleic Acids Res 18: 1648

Venkatesan MM, Buysee JM, Kopecko DJ (1988) Characterization of invasion plasmid antigen genes (*ipaBCD*) from *Shigella flexneri*. Proc Natl Acad Sci USA 85: 9317–9321

Venkatesan MM, Buysse JM, Oaks EV (1992) Surface presentation of *Shigella flexneri* invasion plasmid antigens requires the products of the *spa* locus J Bacteriol 174: 1990–2001

Wassef JW, Keren DF, Mailloux JL (1989) Role of M cells in initial antigen uptake and in ulcer formation in the rabbit intestinal loop model of shigellosis. Infect Immun 57: 858–863

Watanabe H, Arakawa E, Ito KI, Kato JI, Nakamura A (1990) Genetic analysis of an invasion region by use of Tn3-*lac* transposon and identification of a second positive regulator gene, *invE*, for cell invasion of *Shigella sonnei*: significant homology of InvE with ParBo f plasmid P1. J Bacteriol 172: 619–629

Wharton M, Spiegel RA, Horan JM, Tauxe RV, Wells JG, Barg N, Herndon J, Meriwether RA, Newton macCormack J, Levine RH (1990) A large outbreak of antibiotic-resistant shigellosis at a mass gathering. J Infect Dis 162: 1324–1328

Woestyn S, Allaoui A, Wattiau P, Cornelis G (1994) yscN, the putative energizer of the *Yersinia* Yop secretion machinery. J Bacteriol 176: 1561–1569

Wu H, Parsons JT (1993) Cortactin, an 80/85-kilodalton pp60^{src} substrate, in a filamentous actin binding protein enriched in the cell cortex. J Cell Biol 120: 1417–1426

Wu H, Reynolds AB, Kanner SB, Vines RR, Parsons JT (1991) Identification and characterization of a novel cytoskeleton associated pp60^{src} substrate. Mol Cel Biol 11: 5113–5124

Zychlinsky A, Fitting C, Cavaillon JM, Sansonetti PJ (1994a) Interleukin-1 is released by macrophages during apoptosis unduced by *Shigella flexneri*. J Clin Invest 94: 1328–1332

Zychlinsky A, Kenny B, Ménard R, Prevost MC, Holland IB, Sansonetti PJ (1994b) IpaB mediates macrophage apotois induced by *Shigella flexneri*. Mol Microbiol 11: 619–627

Zychlinsky A, Prevost MC, Sansonetti PJ (1992) *Shigella flexneri* induces apoptosis in infected macrophages. Nature 358: 167–168

Molecular and Cellular Bases
of *Salmonella* Entry into Host Cells

J.E. GALÁN

1 Introduction

Salmonella spp. are facultative intracellular pathogens capable of causing disease in a great variety of animal species, including human beings (HOOK 1990). Some *Salmonella* serotypes are highly adapted to a specific host (e.g., *S. typhi* and *S. gallinarum* can infect only human beings and poultry, respectively) or preferentially infect one species (e.g., *S. choleraesuis* and *S. dublin* preferentially infect swine and cattle, respectively). In contrast, other serotypes can infect a broad range of hosts (e.g., *S. enteritidis*). The molecular bases for host adaptation are poorly understood. The type of disease caused by these microorganisms depends not only on the *Salmonella* serotype but also on the species and immunological status of the infected host. In human beings, the clinical manifestations of salmonellosis range from severe systemic infection to mild gastroenteritis. A common feature of the pathogenesis of all Salmonellae is their ability to gain access to cells that are normally nonphagocytic. This includes not only the cells of

Department of Molecular Genetics and Microbiology, School of Medicine, State University of New York at Stony Brook, Stony Brook, NY 11794–5222, USA

the intestinal epithelium, these organisms' port of entry, but also other cells that may constitute "safe sites" for Salmonellae at later stages of their pathogenic cycle (TAKEUCHI 1967; LIN et al. 1987; CONLAN and NORTH 1992; DUNLAP et al. 1992; VERJAN et al. 1994). Although the actual mechanisms of *Salmonella* entry are not fully understood, work in a number of laboratories is beginning to provide some insights into this intricate process. In this chapter, the molecular and cellular bases of *Salmonella* entry into non-phagocytic cells will be reviewed. Most of the information available has been derived from the use of in vitro systems. However, there is enough in vivo evidence that indicates that the molecular genetic bases and mechanistic principles derived from these studies are highly relevant to the natural infection process (GALÁN and CURTISS III 1989; JONES and FALKOW 1994).

2 Interaction of *Salmonella* with the Intestinal Epithelium

Salmonella infections are largely initiated after the consumption of contaminated food or water (CHALKER and BLASER 1988; LEVINE et al. 1991; CDC 1993). Ingested bacteria reach the intestinal tract where they interact in a rather intimate and complex manner with the cells that line the intestinal mucosa. As a result of such an interaction, the microorganisms breach the intestinal epithelium to reach the lamina propria (TAKEUCHI and SPRINZ 1967), where they replicate or proceed to deeper tissues presumably carried within nonactivated macrophages. The site(s) through which *Salmonella* breaches the intestinal barrier has not been precisely defined. Available information suggests that the entry site depends on the host as well as on the *Salmonella* serotype (POWELL et al. 1971; CARTER and COLLINS 1974; BROWN et al. 1976; HOHMANN et al. 1978; McGOVERN and SLAVUTIN 1979; REED et al. 1986). It has been often suggested that, in man, the distal part of the small intestine is the port of *Salmonella* entry. However, there is surprisingly little evidence in the literature to support this assertion (ALEEKSEEV et al. 1960). In a number of experimental animal infections, various segments of the intestinal tract have been shown to be compromised early in the infection process (KENT et al. 1966; CARTER and COLLINS 1974; BROWN et al. 1976; REED et al. 1985, 1986), suggesting that *Salmonella* can breach the intestinal barrier at sites other than the small bowel.

The intestinal barrier is composed of a heterogeneous population of cells present in different numbers throughout the various segments of the intestinal epithelium. By far the most abundant type is the absorptive columnar epithelial cell. Other cell types include goblet cells, very abundant in the large intestine, and the M cells, which are often associated with organized collection of lymphoid follicles known as Peyer's patches. Which of these cell types serves as a passageway for *Salmonella* is a matter of some controversy. M cells have often been suggested as the primary invasion site for *Salmonella* spp. This assertion is supported largely by ligated mouse ileal loop infections, in which both *S. typhi* and *S. typhimurium* were shown to preferentially associate with M cells. Such an interaction ultimately led to the destruction of these cells under these experimental conditions (KOHBATA et al. 1986; JONES et al. 1994). The correlation of these findings with the natural infection process, however, is less clear. It has been shown that oral infection of susceptible mouse strains leads to colonization of Peyer's patches, with subsequent proliferation of the organisms at this site (CARTER and COLLINS 1974; HOHMANN et al. 1978). This observation has often been interpreted as an indication that, in these animals, these structures serve as the primary port of entry for *Salmonella*. However, an alternative interpretation is that Peyer's patches may represent a more favorable environment in which *Salmonella* can more readily multiply. Furthermore, the lack of glycocalix on the surface of M cells may make them more readily accessible to the infecting microorganisms. It is clear that *Salmonella* can also breach the intestinal barrier through the absorptive columnar epithelial cells of the small intestine (TAKEUCHI 1967), and it has been argued that perhaps columnar epithelial cells may indeed constitute the main port of *Salmonella* entry since they vastly outnumber M cells (TRIER and MADARA 1988; McCORMICK et al. 1993). It is likely that the relative importance of the different intestinal epithelial cells in *Salmonella* invasion of the host may be largely dependent on the species of the infected host as well as on the *Salmonella* serotype.

3 *Salmonella* Entry into Nonphagocytic Cells: Role of the Host Cell

It has long been recognized that the host cell plays an active role in the *Salmonella* entry process (KIHLSTROM and NILSSON 1977; BUCKHOLM 1984). Early electron microscopic studies of infected animals showed that shortly after coming into contact with the intestinal epithelium, *Salmonella typhimurium* induced profound changes in the plasma membrane of the infected cells (Takeuchi 1967). These changes, which immediately preceded the uptake of *Salmonella* into a membrane-bound compartment, were characterized by a transient and localized distortion of the brush border of the columnar epithelium of the small intestine. The disruption of the host-cell plasma membrane is accompanied by a significant rearrangement of the cytoskeleton and the accumulation of a number of cytoskeletal-associated proteins such as actin, α-actinin, talin, and ezrin (FINLAY and RUSCHKOWSKI 1991). Cytochalasin D, a drug that disrupts the actin cytoskeleton, effectively prevents bacterial entry, an indication that the bacterial-induced cytoskeletal rearrangements are essential for the internalization process (KIHLSTROM and NILSSON 1977; BUCKHOLM 1984). It is now recognized that the *Salmonella*-induced changes in the plasma membrane resemble the membrane ruffling events triggered by a variety of agonists such as growth factors or by the activation of a number of oncogenes (BAR-SAGI and FERASMICO 1986; KADOWAKI et al.

1986). In addition to the cytoskeletal changes, contact of *Salmonella* with the host cell induces a marked increase in the concentration of intracellular free calcium (GINOCCHIO et al. 1992; PACE et al. 1993) and macropinocytosis (GARCIA-DEL PORTILLO and FINLAY 1994). Interestingly, *Salmonella* triggers calcium fluxes from the outside since addition of cytochalasin D, which effectively prevents bacterial entry, does not impede this response. The ability of *Salmonella* to induce membrane ruffling, calcium fluxes, and macropinocytosis closely correlates with bacterial entry, since invasion-defective mutants were unable to induce these responses in the host cell (GALÁN et al. 1992b; GINOCCHIO et al. 1992; GARCIA-DEL PORTILLO and FINLAY 1994).

It is increasingly clear that *Salmonella* entry into cells is the result of complex signal transduction pathways triggered by the bacteria at the host cell surface (GALÁN 1994). Although the understanding of the signaling events is clearly incomplete, recent studies have begun to provide some details of the cellular processes responsible for bacterial uptake (GALÁN et al. 1992b; RUSCHKOWSKI et al. 1992; JONES et al. 1993; PACE et al. 1993; CHEN et al. 1995) *Salmonella* can trigger membrane ruffling and micropinocytosis in virtually any eukaryotic cell. However, the signaling events that lead to these responses are clearly different in different cells, an observation which has contributed to the apparently contradictory nature of some of the findings (GALÁN 1994). For example, *S. typhimurium* interaction with cultured Henle-407 cells is accompanied by the tyrosine phosphorylation of the epidermal growth factor receptor (EGFR) (GALÁN et al. 1992) and the subsequent stimulation of a complex signaling cascade (PACE et al. 1993). This cascade includes the activation of the mitogen-activated protein (MAP) kinase and one of its downstream substrates, phospholipase A_2 (PLA_2). The activity of the latter enzyme leads to the production of arachidonic acid, which is subsequently converted to a variety of eicosanoids by the action of several enzymes including 5-lipooxygenase (5-LO). One of the eicosanoids, leukotriene D_4, is involved in triggering the calcium fluxes required for *Salmonella* entry. Interruption of this pathway by blocking the activities of PLA_2 or 5-LO effectively prevented *Salmonella* entry. Furthermore, the invasion phenotype of a *Salmonella*-entry defective mutant that was unable to trigger this signaling pathway was effectively rescued by the addition to the infected cells of either EGF, the natural ligand of the EGFR, or LTD_4 (GALÁN et al. 1992b; PACE et al. 1993). *Salmonella* can enter into cells that do not express the EGFR (GALÁN et al. 1992b; FRANCIS et al. 1993), and in those cells the bacteria-induced signaling pathways are different (CHEN et al. 1995). For example, *Salmonella* infection of human epithelioid HeLa cells and mouse B82 fibroblasts, which do not express the EGFR, results in the production of inositol phospholipids [most likely inositol (1, 4, 5) triphosphate], which may ultimately mediate the observed calcium fluxes in these cells (CHEN et al. 1995). Similar pathways are elicited in HeLa cells (RUSCHKOWSKI et al. 1992; CHEN et al. 1995). In contrast, *Salmonella* infection of Henle-407 cells does not result in the production of inositol phospholipids (CHEN et al. 1995).

It is likely that common downstream effector molecules may be involved in the *Salmonella* induced signaling pathway in all cells, since in all cases the ultimate outcome of the *Salmonella*-induced responses is membrane ruffling and cytoskeletal rearrangements that lead to bacterial uptake. Indeed, common signaling components have been identified in various cells. Wild-type *Salmonella* (but not invasion-defective mutants) induce Ca^{2+} mobilization and MAP kinase activation in several cell lines, thereby establishing a correlation between these responses and bacterial entry (GINOCCHIO et al. 1992; PACE et al. 1993; CHEN et al. 1995). Host-cell tyrosine phosphorylation is also required for *Salmonella* entry, since broad-specificity tyrosine kinase inhibitors block bacterial entry into various cell lines (CHEN et al. 1995). Interestingly, a very specific inhibitor of the EGFR tyrosine kinase activity (BUCHDUNGER et al. 1994) blocked entry into Henle-407 cells but not into HeLa or B82 cells, further supporting the notion of alternative signaling pathways in different cells and the involvement of the EGFR pathway in Henle-407 cells (CHEN et al. 1995). ROSENSHINE et al. (1994) have reported that tyrosine kinase inhibitors do not block bacterial entry and that an invasion-defective mutant of *S. typhimurium* was also able to activate MAP kinase. In addition, these authors failed to detect tyrosine phosphorylation of the EGF receptor in *Salmonella*-infected Henle-407 cells. The reasons for these contradictory findings are unclear at the present time but may be related to differences in the experimental protocols, since the *Salmonella* strains and cell lines used in these studies are presumably the same.

It is not understood how these different second messengers mediate bacterial entry. Ca^{2+} is an important regulator of cytoskeletal dynamics, and arachidonic acid metabolites have been shown to regulate the activity of actin-organizing small GTP-binding proteins (TSAI et al. 1990; HAN et al. 1991; CHUAN et al. 1993). However, JONES et al. (1993) showed that microinjection of dominant negative mutants of ras, rac, or the exoenzyme C3 ADP ribosyl transferase, which inhibits the function of rho, did not prevent *Salmonella*-induced membrane ruffling in infected HEp2, Swiss 3T3, and MDCK cells. Although MAP kinase (p^{41mapk}) has been shown to localize to membrane ruffles induced by growth factor stimulation (GONZALEZ et al. 1993), its role in bacterial uptake has not been established. It is possible that this enzyme may play an important role in *Salmonella*-induced nuclear responses that lead to the production of pro-inflammatory cytokines in the infected cells. More studies will be required to elucidate the complexities of the host-cell responses to *Salmonella*.

4 *Salmonella* Entry into Nonphagocytic Cells: Role of the Bacterium

The complexity of the molecular genetic bases of *Salmonella* entry is reflected in the large number of genetic loci involved in this process that have been identified (FINLAY et al. 1988b; ELSINGHORST et al. 1989; GALÁN and CURTISS III 1989; GAHRING et al. 1990; BETTS and FINLAY 1992; GALÁN et al. 1992a; GINOCCHIO et al. 1992; LEE et al. 1992; STONE et al. 1992; ALTMEYER et al. 1993; BEHLAU and MILLER 1993; GROISMAN and OCHMAN 1993; RUBINO et al. 1993; EICHELBERG et al. 1994; JONES and FALKOW 1994; KANIGA et al. 1994; COLLAZO et al.1995). Interestingly, many of these loci are clustered in the 59-min region of the *Salmonella* chromosome. However, additional loci scattered around the chromosome, some of them encoding well-characterized surface determinants, have been implicated in the entry process.

4.1 Determinants of Entry Encoded in the 59-min Region of the *Salmonella* Chromosome

There is mounting evidence indicating that as many as 45 kb encompassing the 59-min region of the *Salmonella* chromosome encode entry determinants (MILLS et al. 1995). This region corresponds to one of the nine "loops" of the *Salmonella* chromosome that share no homology to the *Escherichia coli* K12 chromosome (RILEY and SANDERSON 1990). The G+C content of the genes so far sequenced from this region is very low in comparison to the overall average of the *Salmonella* genome (Galán et al. 1992a; GINOCCHIO et al. 1992; ALTMEYER et al. 1993; GROISMAN and OCHMAN 1993; EICHELBERG et al. 1994; KANIGA et al. 1994). It has therefore been proposed that this segment of the chromosome may have been acquired after horizontal transfer from another organism (GALÁN et al. 1992a; GINOCCHIO et al. 1992). The presence of an IS3-like element in the vicinity of the invasion region of several *Salmonella* serotypes supports this notion (ALTMEYER et al. 1993; GINOCCHIO et al. 1995). In fact, the 59-min region of the chromosome remains unstable in certain *Salmonella* serotypes such as *S. lichtfield* and *S. seftenberg*, giving rise to naturally occurring nonpathogenic strains that carry deletions in this segment (GINOCCHIO et al. 1995). Therefore, the 59-min region of the *Salmonella* chromosome represents another example of a "pathogenicity island".

4.1.1 The *inv/spa* Loci

The molecular characterization of a *S. typhimurium* entry-defective mutant led to the identification of a genetic locus called *inv* (GALÁN and CURTISS III 1989). This locus is required not only for entry into cultured cells but also for *Salmonella* virulence. An *S. typhimurium* strain carrying a polar mutation in *invA* was defective in colonizing the intestinal epithelium of orally infected Balb/c mice. This colonization defect resulted in a higher oral LD_{50} for the Inv⁻ mutant (GALÁN and CURTISS III 1989). The *inv* locus is present in essentially all virulent *Salmonella* (RAHN et al. 1992). Introduction of an *inv* mutation in several *Salmonella* strains belonging to various serotypes rendered them severely defective for entry, indicating not only that this locus is present, but also that it is functional in most, if not all, *Salmonella* spp. (GALÁN and CURTISS III 1991). Subsequent nucleotide sequence analysis of this and adjacent loci identified 13 genes arranged in the following order: *invH, invF, invG, invE, invA, invB, invC, invI, invJ, spaO, spaP, spaR*, and *spaS* (GALÁN et al. 1992a; GINOCCHIO et al. 1992; ALTMEYER et al. 1993; GROISMAN and OCHMAN 1993; EICHELBERG et al. 1994; KANIGA et al. 1994; COLLAZO et al. 1995) (Fig. 1). Functional analysis of nonpolar mutants established that, with the exception of *invB*, all of these genes are required for *S. typhimurium* entry into cultured epithelial cells (GALÁN et al. 1992a; GINOCCHIO et al. 1992; EICHELBERG et al. 1994; KANIGA et al. 1994; COLLAZO and GALÁN 1995; COLLAZO et al. 1995). Furthermore, only mutations in *invH* (ALTMEYER et al. 1993) had any measurable effect on the ability of *Salmonella* to attach to cultured cells, a clear indication that attachment and entry are independent events for this microorganism. Interestingly, the phenotype of *invH* mutants was dependent on the *Salmonella* serotype. The absence of a functional *invH* caused a more pronounced defect in attachment and invasion in the host-adapted *Salmonella* serotypes such as *S. typhi* and *S. gallinarum* (ALTMEYER et al. 1993). The implications of this observation of host specificity are unknown.

Sequence homology of the predicted gene products revealed that InvG, InvA, InvC, InvE, SpaO, SpaP, SpaQ, SpaR, and SpaS are similar to components of what are now considered type-III protein secretion systems (SALMOND and REEVES 1993) (Table 1). This type of protein export system has been identified in a number of bacteria that are pathogenic for plants and animals, including species of the genera *Shigella*, *Yersinia*, *E. coli*, *Pseudomonas*, *Erwinia*, and *Xanthomonas* (SALMOND and REEVES 1993; VAN GIJSEGEM et al. 1993). Some of the proteins that

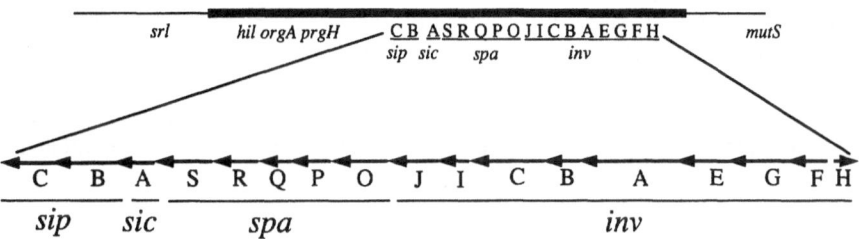

Fig. 1. The 59-min invasion region of the *Salmonella* chromosome

Table 1. Protein homologues in Type III secretion systems in various mammalian and plant pathogens

Salmonella spp.	Shigella spp.	Yersinia spp.	Plant pathogens	Flagellar assembly
InvA	MxiA	LcrD	HrpO	FlhA
InvE	MxiC	LcrE	?	–
InvG	MxiD	YscC	HrpA	–
InvF	MxiE	VirF	HrpB	–
?	MxiJ	YscJ	HrpI	–
InvB	Spa15	?	?	FliH
InvC	Spa47	YscN	HrpE	FliI
SpaO	Spa33	YscP	HrpQ	FliN
SpaP	Spa24	YscQ	HrpT	FliP
SpaQ	Spa9	YscR	MopD	FliQ
SpaR	Spa29	YscS	MopE	FliR
SpaS	Spa40	?	HrpN	–

make up this system are also homologous to components of the flagellar assembly apparatus of gram-positive and gram-negative organisms (DREYFUS et al. 1993). A distinct feature of this type of secretion system is that the target proteins lack typical signal sequences, and therefore this translocon directs export of the target proteins through both the inner and outer membranes of the gram-negative cell wall. Although virtually nothing is known about the mechanisms of the *Salmonella* type-III protein export apparatus, protein sequence and biochemical analysis of the putative components are beginning to yield interesting information. For example, InvA's secondary structure is consistent with that of a polytopic inner-membrane protein with two domains: a hydrophobic amino-terminal domain with at least eight putative transmembrane segments and a hydrophilic carboxyterminal domain, most likely facing the cytoplasm (GALÁN et al. 1992a). Fractionation as well as topological studies using the transposon TnphoA have confirmed this arrangement (GINOCCHIO 1994). This architecture suggests that InvA may be part of a channel that would allow the secreted proteins to get through the inner membrane. However, there is no evidence yet to support this hypothesis. Other putative inner-membrane proteins include SpaP, SpaQ, SpaR, and SpaS since their predicted sequences indicate the presence of internal hydrophobic and presumably transmembrane segments. These proteins, in conjunction with InvA, may constitute an inner-membrane platform where the target proteins are initially engaged by the *inv/spa* translocon. Based on its homology to the PulD family of protein translocases, InvG is predicted to be localized in the outer membrane, where it may serve as a pore for the passage of the target proteins through that membrane (KANIGA et al. 1994). InvC is a member of a family of proteins with homology to the β subunit of the F_oF_1 ATPase family of proteins. Purified InvC exhibited significant ATPase activity. However, an InvC mutant protein with a change in a single amino acid located in the nucleotide-binding region lacked ATPase activity and, when expressed in an *invC* mutant of S. *typhimurium*, failed to complement the invasion phenotype of this strain (EICHELBERG et al. 1994). These results indicate that InvC may serve as the energizer of the translocon encoded by the *inv* and *spa* loci.

The functional similarity among the type-III secretion systems in the different organisms is underscored by the observation that in some instances functional complementation between homologous genes has been observed. For example, mutations in the S. *typhimurium invA* (GINOCCHIO and GALÁN 1995) and *spaP* (GROISMAN et al. 1993) genes were complemented by the cognate *Shigella flexneri* genes *mxiA* and *spa24*, respectively. These interspecies complementation studies have also shown that these systems are tailored for the export of the specific determinants of each pathogen. For example, the *Yersinia enterocolitica lcrD* gene product failed to complement a S. *typhimurium invA* mutation (GINOCCHIO and GALÁN 1995). However, a chimeric protein consisting of the N-terminal half of LcrD and the C-terminal half of InvA successfully complemented an *invA* mutation, indicating that determinants of specificity may lie in the C terminus of this family of proteins.

4.1.2 Targets of the *inv/spa* Translocon

The identification of targets of the *inv/spa*-encoded secretion system is of great interest since it may lead to the identification of effector molecules involved in triggering the internalization process in the target eukaryotic cell. Despite the homology among the components of the type-III secretion system, little homology has been observed among the protein targets identified so far such as the *Yersinia* outer proteins (Yops) (reviewed in STRALEY et al. 1993), the *Shigella* Ipas (reviewed in SANSONETTI 1992), and the harpins of plant pathogenic bacteria (WEI et al. 1992; HE et al. 1993). Recently, a number of proteins whose export depends on the function of the *inv/spa* translocon have been identified (COLLAZO et al. 1995; KANIGA et al. 1995). At least nine proteins were found in culture supernatant of invasion-competent wild-type *Salmonella typhimurium* that were absent from those of several *inv* and *spa* mutants including *invC* and *invG* (KANIGA et al. 1995). One of these proteins, InvJ, is encoded in the *inv* locus and shares no significant homology with any of the identified targets of the related export systems in *Shigella* and *Yersinia* (COLLAZO et al. 1995). However, InvJ has homology, albeit low, to EaeB, a protein from enteropathogenic *E. coli* that has been implicated in triggering the host cell responses that lead to the characteristic rearrangement of the cytoskeleton observed in cells infected with these organisms (DONNENBERG et al. 1993). Interestingly, this protein is also a target of a type-III export system (see Donnenberg this volume). Two additional targets of the *inv/spa* translocon have recently been identified (KANIGA et al. 1995). These proteins, termed SipB and SipC, are also encoded in the 59-min region of the *Salmonella* chromosome and share significant similarity to the *Shigella* IpaB and IpaC proteins, which are required for the complex interactions of *Shigella* with the host cells (SANSONETTI 1992). The similarity between putative *Shigella* and *Salmonella* effector proteins is intriguing, since these organisms interact with the host cells in a different manner. For example, unlike *Salmonella*, (FINLAY et al. 1989), *Shigella* can not readily enter into polarized epithelial cells through the apical side (MOUNIER et al. 1992). Furthermore, *Salmonella* remains in a membrane-bound compartment throughout its intracellular life cycle (TAKEUCHI 1967), in contrast to *Shigella*, which exits from the endocytic vacuole immediately after internalization (SANSONETTI et al. 1986). Determinants of specificity may lie in less conserved regions of these putative effector molecules. This is certainly possible, since the overall identity between SipB and SipC and the *Shigella* homologues IpaB and IpaC is 28% and 25%, respectively, although in certain regions (e.g., a central 150-amino acid hydrophobic domain of SipB and IpaB) the identity is as high as 65%. An alternative explanation is that these proteins may not have effector functions, but may rather facilitate the action or delivery of yet unidentified effector proteins.

It is becoming increasingly clear that targets of the type-III secretion system are associated with specific chaperones in the cytoplasm. Thus chaperones for the Yop proteins of *Yersinia* and the Ipa proteins of *Shigella* have recently been described (Wattiau and Cornelis 1993; Ménard et al. 1994; Wattiau et al. 1994). Two genes encoding putative chaperonin proteins have been identified immediately adjacent to the genes that encode the *Salmonella* exported proteins. One of them, *invI*, encodes a protein of 17 kD with a very high isoelectric point (9.6) and a high probability of forming coiled coils, all features consistent with putative chaperone activity (Collazo et al. 1995). Furthermore, mutations in *invI* rendered *Salmonella* deficient for entry into cultured epithelial cells. The other gene, termed *sicA*, is located immediately upstream of the *sipB* and *sipC* genes and encodes a protein homologous to the *Shigella* and *Yersinia* chaperones IpgC and LcrH (Kaniga et al. 1995). *sicA* is required for *Salmonella* entry, since a nonpolar mutation in this gene rendered *Salmonella* severely deficient for entry.

4.1.3 Contact-dependent Assembly of an Invasion Organelle

High-resolution, low-voltage scanning electron microscopic studies of the interaction of *S. typhimurium* with cultured epithelial cells revealed the presence of appendage-like structures on the surface of *Salmonella* (Ginocchio et al. 1994). These structures, termed invasomes, were seen shortly after the organisms came in close contact with the host cell but were absent from bacteria grown in L-broth under conditions that render these organisms invasion competent. They are approximately 60 nm in diameter and 0.3–1 μm in length. The invasomes seemed to be shed upon triggering of the signaling events that lead to bacterial uptake, since bacteria associated with membrane ruffles lacked the surface appendages. Addition of chloramphenicol immediately before infection did not prevent their assembly, indicating that these structures are made of a pool of presynthesized components. This is consistent with the observation that addition of chloramphenicol does not block the entry of invasion-competent *Salmonella* (Macbeth and Lee 1993). The transient assembly of the invasomes is dependent on the function of the *inv*-encoded translocon since mutations in *invG* and *invC* effectively prevented the contact-dependent assembly of this structure. Interestingly, *invE* mutants were capable of forming the invasome upon contact with host cells, although the appendages looked longer than those observed in wild-type and appeared to have a longer half-life. The putative regulatory role for InvE is consistent with its homology to LcrE (YopN), a *Yersinia* spp. protein involved in the regulation of Yop secretion (Forsberg et al. 1991; Ginocchio et al. 1992).

Although the nature of the invasome appendages has not yet been defined, the requirement of the *inv*-encoded protein export system for its assembly suggests that the targets of this system, such as InvJ, SipB, and SipC, may be components of this structure. Consistent with this notion is the recent observation that contact of wild-type *S. typhimurium* with cultured Henle-407 cells stimulates the secretion of InvJ (Zierler and Galán 1995). Stimulation of InvJ secretion requires live cells, since Henle-407 cells fixed with 2.5% glutaraldehyde failed to stimulate InvJ secretion. Addition of chloramphenicol had virtually no effect on the amount of InvJ released upon contact with Henle-407 cells, indicating that the stimulation of InvJ secretion does not require de novo protein synthesis. Thus, InvJ appears to be stockpiled in *Salmonella* and is subsequently released into the medium during the interaction with the host cells. The secretion of InvJ closely correlates with the invasome assembly process. *invC* and *invG* mutants which failed to assemble the invasome also failed to secrete InvJ upon contact with live Henle-407 cells. In contrast, an *invE* mutant which was capable of assembling the surface appendages (albeit morphologically aberrant) secreted InvJ upon contact with cultured cells at levels equivalent to those of wild type. These results indicate that the presence of Henle-407 cells is capable of activating the type-III secretion system encoded in the *inv* locus, further supporting the notion that *Salmonella* entry into cultured cells is the result of biochemical cross-talk between the bacteria and the host cells.

4.1.4 The *hil, prgH*, and *orgA* Loci

The *Salmonella* entry phenotype is highly regulated (see below). For example, conditions known to affect the degree of DNA superhelicity such as osmolarity (GALÁN and CURTISS III 1990; TARTERA and METCALF 1993), oxygen tension, or growth state (ERNST et al. 1990; LEE and FALKOW 1990; SCHIEMANN and SHOPE 1991) strongly influence bacterial invasion. Therefore, investigators have taken advantage of these regulatory cues to identify genetic loci that are involved in the invasion phenotype. For example, LEE et al. (1992) have isolated *S. typhimurium* mutants capable of entering into cultured epithelial cells when grown under nonpermissive conditions. This strategy identified three classes of mutants. One class of mutants affected a number of *che* genes, thereby causing a smooth-swimming phenotype. This phenotype presumably increases the probability of productive contact between *Salmonella* and the target host cell, thereby aiding the entry process. The second class of mutants affected the expression of *rho*, a transcriptional termination factor. The absence of *rho* may affect the expression of genes encoding invasion determinants. The third class identified a locus termed *hil* (for hyperinvasive locus). Interestingly, this locus was mapped at minute 59 on the *Salmonella* chromosome, approximately 15–20 kb away from the *inv/spa* loci. A deletion encompassing this locus rendered *S. typhimurium* deficient for entry, indicating that it may encode or regulate the expression of invasion determinants that are rate limiting for invasion when the organisms are grown under non-permissive conditions. A different approach was taken by JONES and FALKOW (1994) to identify invasion loci by first identifying oxygen-regulated genes, since growth of *Salmonella* in a low-oxygen environment increases the ability of these organisms to enter mammalian cells. This approach identified *orgA*, which is closely linked to the *hil* locus (JONES and FALKOW 1994). An *S. typhimurium orgA* mutant was reduced in virulence when administered orally to Balb/c mice, although it was fully virulent when administered intraperitoneally, a phenotype very similar to that of an *invA* mutant (GALÁN and CURTISS III 1989). The search for genes whose expression is negatively regulated by the *phoP/phoQ* two-component regulatory system led to the identification of *prgH*, a locus required for *S. typhimurium* entry into cultured cells also very closely linked to *hil* (BEHLAU and

MILLER 1993). Further studies will be necessary to establish the functional relationship among these three as well as the other invasion loci encoded in the 59-min region. The close proximity of these genes on the chromosome, as well as the phenotype of the mutant strains, suggests that all these genes are functionally related.

4.2 Other Entry Loci

Various approaches have identified a number of genetic loci required for entry which are located outside the 59-min region, some of which encode well-characterized surface determinants. It is not yet known whether some (or all) of these loci encode alternative entry pathways, or whether they are somehow functionally related to the invasion functions encoded in the 59-min region. Several studies have demonstrated the importance of flagella (LIU et al. 1988; KHORAMIAN-FALSAFI et al. 1990; JONES et al. 1992; LEE et al. 1992), lipopolysaccharide (LPS) (JONES et al. 1981; FINLAY et al. 1988b; MROCZENSKI-WILDEY et al. 1989) and type-1 pili (ERNST et al. 1990) in the entry process. For example, mutations in *che* genes which conferred the "smooth-swimming" phenotype rendered *S. typhimurium* more invasive than wild type. Mutations in *cheB*, which conferred the "tumbly" phenotype, resulted in strains that were deficient in entry (LIU et al. 1988; KHORAMIAN-FALSAFI et al. 1990; JONES et al. 1992). These observations suggest that the role of motility in bacterial entry may simply be to facilitate productive contact of the bacteria with the target cell. This is further supported by the observation that the invasion defect in nonmotile bacteria can be reversed by applying a mild centrifugal force (JONES et al. 1992) and that Fla⁻ bacteria are fully virulent when administered orally to Balb/c mice (LOCKMAN and CURTISS III 1990). However, it has been reported that in *S. typhi* the invasion defect of Fla⁻ strains cannot be reversed by applying centrifugation, indicating that perhaps in this particular serotype motility and entry are functionally linked in some additional way (LIU et al. 1988).

The role of LPS in entry is also dependent on the *Salmonella* serotype. Rough mutants of *S. choleraesuis* (FINLAY et al. 1988b) and *S. typhi* (MROCZENSKI-WILDEY et al. 1989) are defective for entry, while rough mutants of *S. typhimurium* are not (KIHLSTROM and EDEBO 1976) and rough mutants of *S. enteritidis* are only slightly defective for entry (STONE et al. 1992). It is not known whether LPS plays some role in evoking signaling responses related to invasion, or whether this molecule may simply facilitate contact with the host cells by altering the surface properties of the bacteria.

A variety of transposon mutagenesis analyses have identified several additional *Salmonella* loci required for entry. STONE et al. (1992) isolated a number of Tn*phoA* insertion mutants in *S. enteritidis* that affected invasion to varying degrees. These mutants were mapped to nine different loci on the *Salmonella* chromosome and all but one of them were located outside the 59-min region. Some mutants showed different phenotypes depending on the cell line used to assess their invasiveness. These findings argue for the existence of alternative entry mechanisms in *Salmonella*. FINLAY et al. (1988b) isolated six classes of *TnphoA* insertion mutants in *S. choleraesuis* that were defective in entry. Two of these classes of mutants affected LPS structure, although the remaining four did not. Two of the mutant classes with normal LPS structure were avirulent in mice. Additional noninvasive mutants have been isolated in *S. typhimurium* but not characterized further (BETTS and FINLAY 1992).

ELSINGHORST et al. (1989) isolated a region of the *S. typhi* chromosome that, when introduced into a normally noninvasive strain of *E. coli*, conferred upon this strain low levels of invasiveness. Further analysis identified at least four separate loci required for invasion. The homologous region from the *S. typhimurium* chromosome failed to confer invasive properties to *E. coli*. These invasion loci are linked to the *recA* and sorbitol utilization genes and are therefore adjacent to the 59-min invasion region. The functional significance of this topological relationship is uncertain, since these invasion loci are not part of the 59-min "pathogenicity island" that encodes invasion functions.

5 Regulation of the Invasion Phenotype

The invasion phenotype of *Salmonella* is highly regulated. As described above, many investigators have taken advantage of the different regulatory cues to isolate invasion loci. There are several environmental signals that influence the invasion phenotype. Some of them, such as osmolarity, oxygen tension, or growth state, have in common the fact that they alter the level of DNA superhelicity (HIGGINS et al. 1988; PRUSS and DRLICA 1989; DORMAN et al. 1990). This is consistent with the observation that *Salmonella* entry is strongly influenced by the degree of DNA supercoiling. For example, mutations in *topA*, which drastically alter the level of DNA superhelicity, severely affected the ability of *S. typhimurium* to gain access to host cells (GALÁN and CURTISS III 1990). The *hil* locus, which was isolated by identifying mutants able to enter under nonpermissive growth conditions, is likely to exert its regulatory functions based on these cues (LEE et al. 1992). The *phoP/phoQ* locus is also involved in the regulation of the invasion phenotype since at least one invasion locus, *prgH*, is negatively regulated by this two-component system (BEHLAU and MILLER 1993). This is an interesting finding, as it suggests a regulatory link between two independent, yet related events—bacterial entry and intracellular survival. An additional regulatory mechanism may involve InvF, a member of the *inv* locus homologous to the AraC family of transcriptional regulators. *invF* is essential for *Salmonella* entry into cultured epithelial cells (KANIGA et al. 1994). However, the regulatory targets of this gene remain to be identified. Members of the AraC family of transcription regulators have also been implicated in the regulation of related phenotypes such as *Shigella* invasion (SAKAI et al. 1988) or the secretion and production of the *Yersinia* Yops (CORNELIS et al. 1989). Another regulatory layer includes the mechanisms that orchestrate the careful assembly of the surface organelle or invasome that is required for *Salmonella* entry (GINOCCHIO et al. 1994). Although such regulatory mechanisms are poorly understood, it is known that the regulatory cue(s) is provided by live eukaryotic cells (see previous section) (ZIERLER and GALÁN 1995) and that the regulatory effect is not exerted at the transcription or translational level. More work will need to be done in order to understand the complex interaction between these various regulatory networks.

6 Concluding Remarks

We now know that *Salmonella* entry into host cells is the result of true "cross-talk" between the bacterium and the host cell. It is clear that bacteria can trigger within the host cell a complex signaling cascade that leads to membrane ruffling, micropinocytosis, and subsequent bacterial uptake. It is also apparent that the presence of the host cell activates a novel, sec-independent dedicated protein secretion system with the subsequent assembly of appendage-like structures on the surface of the bacterium. The challenge for the next few years will be to identify the actual molecules that mediate this remarkable cross-talk. As we understand more about how *Salmonella* enters into host cells, we will undoubtedly learn more about protein secretion, organelle assembly, and even basic cellular processes such as transmembrane signaling.

Acknowledgments. I would like to thank Katrin Eichelberg and Stephanie Tucker for careful review of this manuscript. Work in my laboratory is supported by grants from the National Institutes of Health and the American Heart Association.

References

Aleekseev PA, Berman MI, Korneeva EP (1960) Clinical and histological picture of *Salmonella typhimurium* infection in children. J Microbiol Epidemiol Immunobiol 31: 133–139

Altmeyer RM, McNern JK, Bossio JC, Rosenshine I, Finlay BB, Galán JE (1993) Cloning and molecular characterization of a gene involved in *Salmonella* adherence and invasion of cultured epithelial cells. Mol Microbiol 7: 89–98

Bar-Sagi D, Ferasmico JR (1986) Induction of membrane ruffling and fluid-phase pinocytosis in quiescent fibroblasts by the ras proteins. Science 233: 1061–1068

Behlau I, Miller SJ (1993) A Pho-P-repressed gene promotes *Salmonella typhimurium* invasion of epithelial cells. J Bacteriol 175: 4475–4484

Betts J, Finlay BB (1992) Identification of *Salmonella typhimurium* invasiveness loci. Can. J Microbiol 38: 852–857

Brown DD, Ross JG, Smith AFG (1976) Experimental infections of sheep with *Salmonella typhimurium.* Res Vet Sci 21: 335–340

Buchdunger E, Trinks U, Mett H, Regenass U, Müller M, Meyer T, McGlynn E, Pinna LA, Traxler P, Lydon NB (1994) 4,5-Dianilinophthalimide: a protein-tyrosine inhibitor with selectivity for the epidermal growth factor receptor signal transduction pathway and potent in vivo antitumor activity. Proc Natl Acad Sci USA 91: 2334–2338

Buckholm G (1984) Effect of cytochalasin B and dihidrocytochalasin B on invasiveness of entero-invasive bacteria in Hep-2 cell cultures. Acta pathol Microbiol Immunol Scand 92: 145–149

Carter P, Collins F (1974) The route of enteric infection in normal mice. J Exp Med 139: 1189–1203

CDC (1993) Outbreaks of *Salmonella enteritidis*—California 1993. MMWR 42: 793–797

Chalker RB, Blaser MJ (1988) A review of human salmonellosis. III. Magnitude of *Salmonella* infections in the United States. Rev Infect Dis 10: 111–124

Chen L-M, Pace J, Galán JE (1995) Common components in the signaling pathways triggered by *Salmonella typhimurium* in different cell lines. Infect Immun (submitted)

Chuan T-H, Bohl BP, Bokoch GM (1993) Biologically active lipids are regulators of Rac-GDI complexation. J Biol Chem 268: 26206–26211

Collazo CM, Galán JE (1995) Functional analysis of the *spa* locus of *Salmonella typhimurium*. Infect Immun (submitted)

Collazo CM, Zierler MK, Galán JE (1995) Functional analysis of the *Salmonella typhimurium* invasion genes *invI* and *invJ* and identification of a target of the protein secretion apparatus encoded in the *inv* locus. Mol Microbiol 15: 25–38

Conlan JW, North RJ (1992) Early pathogenesis of infection in the liver with the facultative intracellular bacteria *Listeria monocytogenes*, *Francisella tularensis*, and *Salmonella typhimurium* involves lysis of infected hepatocytes of leukocytes. Infect Immun 60: 5164–5171

Cornelis G, Sluiters C, de Rouvroit CL, Michiels T (1989) Homology between VirF, the transcriptional activator of the *Yersinia* virulence regulon, and AraC, the *Escherichia coli* arabinose operon regulator. J Bacteriol 171: 254–262

Donnenberg MS, Yu J, Kaper JB (1993) A second chromosomal gene necessary for intimate attachment of enteropathogenic *Escherichia coli* to epithelial cells. J Bacteriol 175: 4670–4680

Dorman CJ, Ni Bhriain N, Higgins CF (1990) DNA supercoiling and environmental regulation of virulence gene expression in *Shigella flexneri*. Nature 344: 789–792

Dreyfus G, Williams AW, Kawagishi I, Macnab RM (1993) Genetic and biochemical analysis of *Salmonella typhimurium* FliI, a flagellar protein related to the catalytic subunit of the F_0F_1 ATPase and to virulence proteins of mammalian and plant pathogens. J Bacteriol 175: 3131–3138

Dunlap NE, Benjamin W Jr., Berry AK, Eldridge JH, Briles DE (1992) A `safe-site' for *Salmonella typhimurium* is within splenic polymorphonuclear cells. Microb Pathog 13: 181–190

Eichelberg K, Ginocchio C, Galán JE (1994) Molecular and functional characterization of the *Salmonella typhimurium* invasion genes *invB* and *invC* : homology of *InvC* to the F_0F_1 ATPase family of proteins. J Bacteriol 176: 4501–4510

Elsinghorst EA, Baron LS, Kopecko DJ (1989) Penetration of human intestinal epithelial cells by *Salmonella*: molecular cloning and expression of *Salmonella typhi* invasion determinants in *Escherichia coli*. Proc Natl Acad Sci USA 86: 5173–5177

Ernst RK, Domboski DM, Merrick JM (1990) Anaerobiosis, type 1 fimbriae, and growth phase are factors that affect invasion of Hep-2 cells by *Salmonella typhimurium*. Infect Immun 58: 2014–2016

Finlay BB, Gumbiner B, Falkow W (1988a) Penetration of *Salmonella* through a polarized Madin-Darby canine kidney epithelial cell monolayer. J Cell Biol 107: 221–230

Finlay BB, Starnbach MN, Francis CL, Stocker BAD, Chatfield S, Dougan G, Falkow S (1988b) Identification and characterization of TnphoA mutants of *Salmonella* that are unable to pass through a polarized MDCK epithelial cell monolayer. Mol Microbiol 2: 757–766

Finlay BB, Ruschkowski S (1991) Cytoskeletal rearrangements accompanying *Salmonella* entry into epithelial cells. J Cell Sci 99: 283–296

Forsberg A, Vitanen AM, Skurnik M, Wolf-Watz H (1991) The surface-located YpoN protein is involved in calcicm signal transduction in *Yersinia pseudotuberculosis*. Mol Microbiol 5: 977–986

Francis, CL, Ryan TA, Jones BD, Smith SJ, Falkow S (1993) Ruffles induced by *Salmonella* and other stimuli direct macropinocytosis of bacteria. Nature 364: 639–642

Gahring LC, Heffron F, Finlay BB, Falkow S (1990) Invasion and replication of *Salmonella typhimurium* in animal cells. Infect Immun 58: 443–448

Galán JE (1994) *Salmonella* entry into mammalian cells: different yet converging signal transduction pathways? Trends Cell Bio 4: 196–199

Galán JE, Curtiss R III (1989) Cloning and molecular characterization of genes whose products allow *Salmonella typhimurium* to penetrate tissue culture cells. Proc Natl Acad Sci USA 86: 6383–6387

Galán JE, Curtiss R III (1990) Expression of *Salmonella typhimurium* genes required for invasion is regulated by changes in DNA supercoiling. Infect Immun 58: 1879–1885

Galán JE, Curtiss R III (1991) Distribution of the *invA, -B, -C and -D* genes of *Salmonella typhimurium* among other *Salmonella* serovars: *invA* mutants of *Salmonella typhimurium* are deficient for entry into mammalian cells. Infect Immun 59: 2901–2908

Galán JE, Ginocchio C, Costeas P (1992a) Molecular and functional characterization of the *Salmonella typhimurium* invasion gene *invA*: homology of InvA to members of a new protein family. J Bacteriol 17: 4338–4349

Galán JE, Pace J, Hayman MJ (1992b) Involvement of the epidermal growth factor receptor in the invasion of the epithelial cells by *Salmonella typhimurium*. Nature 357: 588–589

Garcia-del Portillo F, Finlay BB (1994) *Salmonella* invasion of nonphagocytic cells induces formation of macropinosomes in the host cell. Infect Immun 62: 4641–4645

Ginocchio C, Galán JE (1995) Functional conservation among members of the *Salmonella typhimurium* InvA family of proteins. Infect Immun 63: 729–732

Ginocchio C, Olmsted SB, Wells CL, Galán JE (1994) Contact with epithelial cells induces the formation of surface appendages on *Salmonella typhimurium*. Cell 76: 717–724

Ginocchio C, Pace J, Galán JE (1992) Identification and molecular characterization of a *Salmonella typhimurium* gene involved in triggering the internalization of Salmonellae into cultured epithelial cells. Proc Natl Acad Sci USA 89: 5976–5980

Ginocchio C, Rahn K, Clark RC, Galán JE (1995) Naturally occurring deletions in the *inv* locus of environmental isolates of *S. seftenberg* and *S. litchfield*. Infect Immun (submitted)

Gonzalez FA, Seth A, Raden DL, Bowman DS, Fay FS, Davis RJ (1993) Serum-induced translocation of mitogen-activated protein kinase to the cell surface ruffling membrane and the nucleus. J Cell Biol 122: 1089–1101

Groisman EA, Ochman H (1993) Cognate gene clusters govern invasion of host epithelial cells by *Salmonella typhimurium* and *Shigella flexneri*. EMBO J 12: 3779–3787

Groisman EA, Sturmoski MA, Solomon FR, Lin R, Ochman H (1993) Molecular, functional, and evolutionary analysis of sequences specific to *Salmonella*. Proc Natl Acad Sci USA 90: 1033–1037

Han JW, McCormick F, Macara IG (1991) Regulation of Ras-Gap and neurofibromatosis-1 gene product by eicosanoids. Science 252: 576–579

He SY, Huang H-C, Collmer A (1993) *Pseudomonas syringae* pv. *syringae* Harpin[Pss]: a protein that is secreted via the Hrp pathway and elicits the hypersensitive response in plants. Cell 73: 1255–1266

Higgins CF, Dorman CJ, Stirling DA, Waddell L, Booth IR, May G, Bremer E (1988) A physiological role for DNA supercoiling in the osmotic regulation of gene expression in *S. typhimurium* and *E. coli*. Cell 52: 569–584

Hohmann AW, Schmidt G, Rowley D (1978) Intestinal colonization and virulence of *Salmonella* in mice. Infect Immun 22: 763–770

Hook EW (1990) Salmonella species (including typhoid fever). In: Mandell GL(eds)Principles and practice of infectious diseases. Wiley, New York

Jones BD, Falkow S (1994) Identification and characterization of a *Salmonella typhimurium* oxygen-regulated gene required for bacterial internalization. Infect Immun 62: 3745–3752

Jones GW, Richardson LA, Uhlman D (1981) The invasion of HeLa cells by *Salmonella typhimurium*: reversible and irreversible bacterial attachment and the role of bacterial motility. J Gen. Microbiol 127: 351–360

Jones BD, Lee CA, Falkow S (1992) Invasion by *Salmonella typhimurium* is affected by the direction of flagellar rotation. Infect Immun 60: 2475–2480

Jones BD, Paterson HF, Hall A, Falkow S (1993) *Salmonella typhimurium* induces membrane ruffling by a growth factor-receptor-independent mechanism. Proc Natl Acad Sci USA 90: 10390–10394

Jones BD, Ghori N, Falkow S (1994) *Salmonella typhimurium* initiates murine infection by penetrating and destroying the specialized epithelial M cells of the Peyer's patches. J Exp Med 180: 15–23

Kadowaki T, Koyasu S, Nishida E, Sakai H, Takaku F, Yahara I, Kasuga M (1986) Insulin-like growth factors, insulin, and epidermal growth factor cause rapid cytoskeletal reorganization in KB cells. J Biol Chem 261: 16141–16147

Kaniga K, Bossio JC, Galán JE (1994) The *Salmonella typhimurium* invasion genes *invF* and *invG* encode homologues to the PulD and AraC family of proteins. Mol Microbiol 13: 555–568

Kaniga K, Tucker SC, Galán JE (1995) Homologues of the *Shigella* invasions IpaB and IpaC are required for *Salmonella typhimurium* entry into cultured cells. J Bacteriol (in press)

Kent TH, Formal SB, Labrec EH (1966) *Salmonella* gastroenteritis in rhesus monkeys. Arch Pathol 82: 272–279

Khoramian-Falsafi T, Harayama S, Kutsukake K, Pechere JC (1990) Effect of motility and chemotaxis on the invasion of *Salmonella typhimurium* into HeLa cells. Microb Pathog 9: 47–53

Kihlstrom E, Edebo L (1976) Association of viable and inactivated *Salmonella typhimurium* 395 MS and MR 10 with HeLa cells.Infect Immun 14: 851–857

Kihlstrom E, Nilsson L (1977) Endocytosis of *Salmonella typhimurium* 395 MS and MR 10 by HeLa cells.Acta pathol Microbiol Scand 85: 322–328

Kohbata S, Yokoyama H, Yabuuchi E (1986) Cytopathogenic effect of *Salmonella typhi* GIFU 10007 on M cells of murine ileal Peyer's patches in ligated ileal loops: an ultrastructural study. Microbiol Immunol 30: 1225–1237

Lee CA, Falkow S (1990) The ability of *Salmonella* to enter mammalian cells is affected by bacterial growth state.Proc Natl Acad Sci USA 87: 4304–4308

Lee CA, Jones BD, Falkow S (1992) Identification of a *Salmonella typhimurium* invasion locus by selection of hyperinvasive mutants. Proc Natl Acad Sci USA 89: 1847–1851

Levine WC, Buehler JW, Bean NH, Tauxe RV (1991) Epidemiology of nontyphoidal *Salmonella* bacteremia during the human immunodeficiency virus epidemic. J Infect Dis 164: 81–87

Lin F-R, Wang X-M, Hsu HS, Mumaw VR, Nakoneczna I (1987) Electron microscopic studies on the location of bacterial proliferation in the liver in the murine salmonellosis. Br J Exp Pathol 68: 539–550

Liu SL, Ezaki T, Miura H, Matsui K, Yabuuchi E (1988) Intact motility as a *Salmonella typhi* invasion-related factor. Infect Immun 56: 1967–1973

Lockman HA, Curtiss III R (1990) *Salmonella typhimurium* mutants lacking flagella or motility remain virulent in BALB/c mice. Infect Immun 58: 137–143

Macbeth KJ, Lee CA (1993) Prolonged inhibition of bacterial protein synthesis abolishes *Salmonella* invasion. Infect Immun 61: 1544–1546

McCormick BA, Colgan SP, Delp-Archer C, Miller SI, Madara JL (1993) *Salmonella typhimurium* attachment to human intestinal epithelial monolayers: transcellular signalling to subepithelial neutrophils. J Cell Biol 123: 895–907

McGovern VJ, Slavutin LJ (1979) Pathology of *Salmonella* colitis. Am J Surg Pathol 3: 483–490

Ménard R, Sansonetti PJ, Parsot C, Vasselon T (1994) The IpaB and IpaC invasins of *Shigella flexneri* associate in the extracellular medium and are partitioned in the cytoplasm by a specific chaperon. Cell 76: 829–839

Mills DB, Bajaj V, Lee CA (1995) A 40 kilobase chromosomal fragment encoding *Salmonella typhimurium* invasion genes is absent from the corresponding region of the *Escherichia coli* K-12 chromosome. Mol Microbiol (in press)

Mounier J, Vasselon T, Hellio R, Lesourd M, Sansonetti PJ (1992) *Shigella flexneri* enters human colonic Caco-2 epithelial cells through the basolateral pole. Infect Immun 60: 237–248

Mroczenski-Wildey MJ, Di Fabio JL, Cabello FC (1989) Invasion and lysis of HeLa cell monolayers by *Salmonella typhi*: the role of lipopolysaccharide. Microb Pathog 6: 143–152

Pace J, Hayman MJ, Galán JE (1993) Signal transduction and invasion of epithelial cells by *Salmonella typhimurium*. Cell 72: 505–514

Powell DW, Plotkin GR, Maenza RM, Solberg LI, Catlin DH, Formal SB (1971) Experimental diarrhoea I. Intestinal water and electrolyte transport in rat *Salmonella* enterocolitis. Gastroenterology 60: 1053–1063

Pruss GJ, Drlica K (1989) DNA supercoiling and prokaryotic transcription. Cell 56: 521–523

Rahn K, De Grandis S, Clark RC, McEwen SA, Galán JE, Ginocchio C, Curtiss III R, Gyles CL (1992) Amplification of an *invA* gene sequence of *Salmonella typhimurium* by polymerase chain reaction as a specific method of detection of *Salmonella*. Mol Cell Probes 6: 271–279

Reed WM, Olander HJ, Thacker HL (1985) Studies on the pathogenesis of *Salmonella heidelberg* infection in weanling pigs. Am J Vet Res 46: 2300–2310

Reed WM, Olander HJ, Thacker HL (1986) Studies on the pathogenesis of *Salmonella typhimurium* and *Salmonella choleraesuis* var *kunzendorf* infection in weanling pigs. Am J Vet Res 47: 75–83

Riley M, Sanderson KE (1990) Comparative genetics of *Escherichia coli* and *Salmonella typhimurium*. In: Drlica K, Riley M (eds) The bacterial chromosome. American Society for Microbiology, Washington, D.C.

Rosenshine I, Ruschkowski S, Foubister V, Finlay BB (1994) *Salmonella typhimurium* invasion of epithelial cells: role of the induced host cell tyrosine protein phosphorylation. Infect Immun 62: 4969–4974

Rubino S, Leori G, Rizzu P, Erre G, Colombo MM, Uzzau S, Masala G, Cappuccinelli P (1993) Tn*phoA Salmonella abortusovis* mutants unable to adhere to epithelial cells and with reduced virulence in mice. Infect Immun 61: 1786–1792

Ruschkowski S, Rosenshine I, Finlay BB (1992) *Salmonella typhimurium* induces an inositol phosphate flux in infected epithelial cells. FEMS Lett 74: 121–126

Sakai T, Sasakawa C, Yoshikawa M (1988) Expression of four virulence antigens of *Shigella flexneri* is positively regulated at the transcriptional level by the 30 kilodalton *virF* protein. Mol Microbiol 2: 589–597

Salmond GPC, Reeves PJ (1993) Membrane traffic wardens and protein secretion in gram-negative bacteria. Trends Biochem Sc 18: 7–12

Sansonetti PJ (1992) Molecular and cellular biology of *Shigella flexneri* invasiveness: from cell assay systems to shigellosis. Curr Top Microbiol Immunol 180: 1–19

Sansonetti PJ, Ryter A, Clerc P, Maurelli AT, Mounier J (1986) Multiplication of *Shigella flexneri* within HeLa cells: lysis of the phagocytic vacuole and plasmid-mediated contact hemolysis. Infect Immun 51: 461–469

Schiemann DA, Shope SR (1991) Anaerobic growth of *Salmonella typhimurium* results in increased uptake by Henle 407 epithelial and mouse peritoneal cells in vitro and repression of a major outer membrane protein. Infect Immun 59: 437–440

Stone BJ, Garcia CM, Badger JL, Hassett T, Smith RIF, Miller V (1992) Identification of novel loci affecting entry of *Salmonella enteritidis* into eukaryotic cells. J Bacteriol 174: 3945–3952

Straley SC, Skrzypek E, Plano GV, Bliska JB (1993) Yops of *Yersinia* spp. pathogenic for humans. Infect Immun 61: 3105–3110

Takeuchi A (1967) Electron microscopic studies of experimental *Salmonella* infection.1. Penetration into the intestinal epithelium by *Salmonella typhimurium*. Am J Pathol 50: 109–136

Takeuchi A, Sprinz H (1967) Electron-microscope studies of experimental *Salmonella* infection in the preconditioned guinea pig. II. Response of the intestinal mucosa to the invasion by *Salmonella typhimurium*. Am J Pathol. 51: 137–161

Tartera C, Metcalf ES (1993) Osmolarity and growth phase overlap in regulation of *Salmonella typhi* adherence to and invasion of human intestinal cells. Infect Immun 61: 3084–3089

Trier JS, Madara JL (1988) Morphology of the mucosa of the small intestine. In: Johnson LR (ed) Physiology of the gastrointestinal tract, vol 2.Raven, New York

Tsai MH, Yu CL, Stacey DW (1990) A cytoplasmic protein inhibits the GTPase activity of H-Ras in a phospholipid-dependent manner. Science 250: 962–965

Van Gijsegem F, Genin S, Boucher C (1993) Conservation of secretion pathways for pathogenicity determinants of plant and animal bacteria. Trends Microbiol 1: 175–180

Verjan GMM, Ringrose JH, van Alphen L, Feltkamp TEW, Kusters JG (1994) Entrance and survival of *Salmonella typhimurium* and *Yersinia enterocolitica* within human B- and T-cell lines.Infect Immun 62: 2229–2235

Wattiau P, Bernier B, Deslée P, Michiels T, Cornelis GR (1994) Individual chaperones required for Yop secretion by *Yersinia*.Proc Natl Acad Sci USA 91: 10493–10497

Wattiau P, Cornelis GR (1993) SycE, a chaperone-like protein of *Yersinia enterocolitica* involved in the secretion of YopE. Mol Microbiol 8: 123–131

Wei ZM, Laby RJ, Zumoff CH, Bauer DW, He SY, Collmer A, Beer SV (1992) Harpin, elicitor of the hypersensitive response produced by the plant pathogen *Erwinia amylovora*. Science 257: 85–88

Zierler M, Galán JE (1995) Contact with cultured epithelial cells induces the secretion of the *Salmonella typhimurium* invasion protein. Inv J Infect Immun (submitted)

Molecular and Genetic Determinants Involved in Invasion of Mammalian Cells by *Listeria monocytogenes*

S. Dramsi, M. Lebrun, and P. Cossart

1 Introduction

Intracellular pathogens can be artificially divided into two groups, those which are present only in phagocytes and those which also enter and replicate in nonprofessional phagocytic cells. The process of entry into nonphagocytic cells is often referred to as "parasite-directed endocytosis" or "induced phagocytosis", since it involves the participation of both the pathogen and the host eukaryotic cell. In those cases where it has been carefully studied, interaction between a bacterial ligand and the host cell receptor results in stimulation of host signaling pathways and the subsequent cytoskeletal rearrangements necessary for uptake (Bliska et al. 1993). Phagocytosis by professional phagocytes is generally different from "induced phagocytosis" in at least two aspects: (a) Internalization is mainly a "host-directed" event resulting from deposition on the bacterium of antibodies or other compounds recognized by phagocyte receptors. (b) Professional phagocytes can produce potent microbicidal radicals or compounds that intracellular bacteria must circumvent, alter, or destroy in order to survive.

Listeria monocytogenes, a gram-positive facultative intracellular bacterium responsible for severe food-borne infections in humans and animals (Fleming et al. 1985; Gray and Killinger 1966; Schlech III et al. 1983), belongs to the

Unité des Interactions Bactéries-Cellules, CNRS URA 1300, Institut Pasteur, 28 rue du Docteur Roux, 75724 Paris Cedex 15, France

second category of intracellular pathogens described above. During infection, this bacterium is found both in macrophages (MACKANESS 1962, 1964) and in normally nonphagocytic cells, such as intestinal epithelial cells (RACZ et al. 1970, 1972) and hepatocytes (LEPAY et al. 1985; ROSEN et al. 1989). While replication in professional phagocytes has long been appreciated as being critical to the development of listeriosis, we are now beginning to realize that entry and replication in normally nonphagocytic cells is also an important component of bacterial pathogenesis. For example, recent work has demonstrated that bacterial multiplication in hepatocytes plays a key role in the development of a systemic infection (CONLAN and NORTH 1991; see below).

The use of simple in vitro systems of infection (for a review see COSSART and MENGAUD 1989; GAILLARD et al. 1987; KUHN et al. 1988; WOOD et al. 1993) and the development of genetic techniques has greatly facilitated the study of the genetic and molecular determinants of *L. monocytogenes* pathogenesis, including those involved in entry into mammalian cells. Previously, the molecular mechanisms of bacterial entry into nonprofessional phagocytes was studied mainly in gram-negative bacteria, such as *Yersinia, Salmonella,* and *Shigella* spp. (for a review see FALKOW et al. 1992). Based on recent studies of invasion by *L. monocytogenes,* it now seems likely that this bacterium enters host cells by mechanisms that are quite different than those used by these enteric bacteria. This review focuses only on the current understanding of the process of entry of *L. monocytogenes* into mammalian cells. The other steps of the infection process have been extensively reviewed elsewhere (COSSART 1994, 1995; COSSART and KOCKS 1994; PORTNOY 1995; SHEEHAN et al. 1994)

2 The Human and Murine Infections

Listeriosis is a food-borne infection which generally occurs after consumption of contaminated food stuff, suggesting that the main site of entry in the organism is the intestine (FLEMING et al. 1985; SCHLECH et al. 1983). It affects mainly immuno-compromised individuals and pregnant women. The clinical features of the disease include meningitis or meningoencephalitis, septicemia, abortion, and perinatal infections (GRAY and KILLINGER 1966).

Much of our understanding of the disease comes from the mouse model and from classic studies of induction of cell-mediated immunity (for a review see KAUFMANN 1993). These studies revealed that a primary infection with *L. monocytogenes* induces a very efficient T-cell response, and that antibodies play no measurable role in recovery from infection or protection against a secondary infection.

Most aspects of human listeriosis can be experimentally reproduced in the mouse model (AUDURIER et al. 1980; MACKANESS 1962; NORTH 1970, 1973). Mice inoculated orally with high doses of *L. monocytogenes* (10^8–10^9 in BALB/c mice) develop a systemic illness. The primary site of intestinal tissue invasion by *L. monocytogenes* is still the subject of uncertainty. Two sites have been described: enterocytes (Racz et al. 1972) and Peyer's patches (MacDonald and Carter 1980; Marco et al. 1992). In the case of Peyer's patches, *L. monocytogenes* probably penetrates through the M cells, which appear particularly active in internalizing other bacterial pathogens such as *Yersinia, Shigella,* and *Salmonella* (Carter and Collins 1974; Grutzkau et al. 1990; Hanski et al. 1989; Wassef et al. 1989). Following translocation across the intestinal barrier, *L. monocytogenes* can be observed in phagocytic cells present in the underlying lamina propria (Racz et al. 1972). Bacteria then spread via the lymph and the bloodstream to the liver and the spleen.

In the liver, the bacteria are rapidly phagocytosed by the Kupffer cells, the resident macrophages lining the liver sinusoids, and 90% of the inoculum is destroyed during the first 6 h (Lepay et al. 1985; Mackaness 1962). Thereafter, the survivors infect adjacent hepatocytes, which are destroyed by neutrophils during the next 24 h (Conlan and North 1991; Rogers and Unanue 1993; Rosen et al. 1989). Lysed hepatocytes liberate the bacteria, which can then be ingested by neutrophils or macrophages. Thus neutrophils play a key role in the early defense mechanism. The importance of neutrophil recruitment in limiting the infection of hepatocytes by *L. monocytogenes* was shown by treating mice with a monoclonal antibody (5C6) directed against the type-3 complement receptor (CR3), a heterodimeric β2 integrin molecule (CD18–CD11b, also called, $\alpha_{mac}\beta2$) (Conlan and North 1991; Rosen et al. 1989). 5C6 is specific for the CD11b polypeptide of CR3 and inhibits the ability of neutrophils and monocytes to adhere to the intercellular adhesion molecule 1 (ICAM1), which is inducibly expressed on the surface of vascular endothelial cells. In this way, 5C6 inhibits migration of neutrophils from blood to sites of inflammation. In mice treated with 5C6, bacteria multiplied unrestrictedly in the hepatocytes (Rosen et al. 1989).

Subsequently during the infection process, depending on the further immune response of the host, the bacteria are either eliminated or undergo further hematogenous dissemination to the brain or placenta (Gray and Killinger 1966).

3 *Listeria*-Macrophage Interactions

Mackaness was the first to demonstrate that *L. monocytogenes* is able to survive and replicate in murine macrophages (Mackaness 1962, 1964). This property has long been considered a major determinant for the virulence of this organism. It is now clear, however, that other factors are also crucial for the establishment of bacterial infection.

During infection, *L. monocytogenes* encounters various types of macrophages: resident macrophages present beneath M cells, Kupffer cells in the liver, and activated macrophages which are recruited to the site of infection. In vitro studies have shown that mononuclear phagocytes are very heterogeneous with regard to listericidal activity; some macrophages allow intracellular growth of this bacterium, whereas others kill it (ALFORD et al. 1991; BAKER and CAMPBELL 1980; CAMPBELL 1994; DE CHASTELLIER and BERCHE 1994; MACKANESS 1962; MIAKE et al. 1980; SPITALNY 1981). In listericidal macrophages, the bacteria remain trapped in the phagosomal compartment (DREVETS et al. 1992; PORTNOY et al. 1989). Components of the *Listeria* cell wall activate complement by the alternative pathway (CROIZÉ et al. 1993) and, after deposition of C3b on the bacterium, entry into macrophages occurs in a complement-dependent manner. The nature of the receptor used during phagocytosis seems to influence the killing of the phago-cytosed bacteria. Listericidal (proteose peptone-elicited) peritoneal macrophages phagocytose *Listeria* primarily through CR3 (DREVETS and CAMPBELL 1991; DREVETS et al. 1992, 1993), while nonlistericidal peritoneal macrophages (thioglycolate-elicited) appear to phagocytose *Listeria* primarily through receptors other than CR3 (ALVAREZ-DOMINGUEZ et al. 1993; DREVETS et al. 1992). This is interesting, because for several other bacteria, CR3-mediated uptake into phagocytes allows the microorganism to bypass toxic pathways involving production of H_2O_2 and oxygen free radicals and thus is less microbicidal. Phagocytic internalization of particles coated with antibody induces the release of H_2O_2, whereas particles coated with C3bi do not induce H_2O_2 release during phagocytosis (WRIGHT and SILVERSTEIN 1983). Therefore, it seems likely that listericidal macrophages kill *L. monocytogenes* by pathways other than oxygen free radical production, such as those involving NO release (BECKERMAN et al. 1993; BOOCKVAR et al. 1994).

There are other ways in which *Listeria* could enter phagocytic cells. It has been shown that the type-I macrophage scavenger receptor binds to gram-positive bacteria and recognizes lipoteichoic acid (DUNNE et al. 1994). Lipoteichoic acids are ubiquitous gram-positive bacterial cell surface components. Thus, scavenger receptors may participate in host defense by clearing lipoteichoic acids and/or intact bacteria from tissues and the bloodstream. However, the relevance of this phenomenon has not been assessed.

Once internalized by macrophages, the bacteria are engulfed in membrane-bound vacuoles. In listericidal macrophages, the bacteria remain trapped in the phagosomal compartment (DREVETS et al. 1992; PORTNOY et al. 1989), while in nonlistericidal macrophages the bacteria are able to escape from this vacuole. Thus, another factor that may play an important role in determining the listericidal activity of macrophages is the ability to retain *L. monocytogenes* in the phagosome.

Bacterial escape from the vacuole and further growth in the cytoplasm depend on the key virulence factor listeriolysin O (see SHEEHAN et al. 1994). Therefore, the apparent replication in macrophages is a reflection of not only intracytosolic multiplication but also intracellular killing in the acidified phagosome (DE CHASTELLIER and BERCHE 1994). The following steps of the infection process in macrophages were indentified in vitro using the murine macrophage cell line J774 (TILNEY and PORTNOY 1989) and are summarized in Fig. 1. The main result of this study was the discovery of a mechanism allowing the direct cell-to-cell

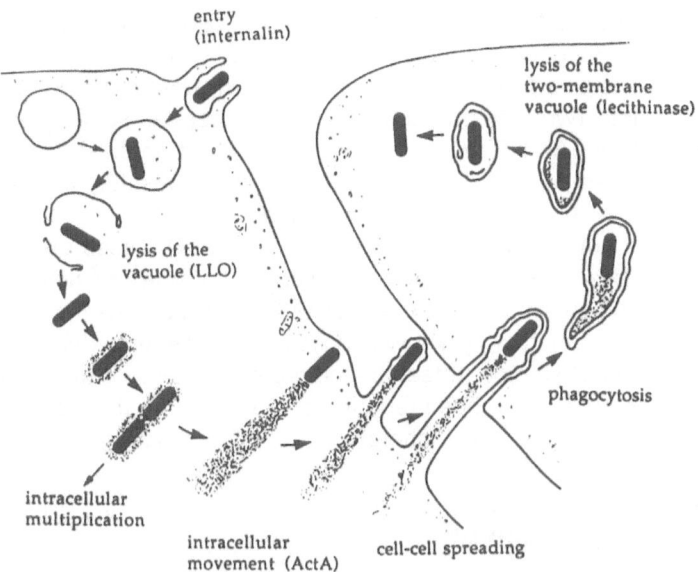

entry
(internalin)

lysis of the
two-membrane
vacuole (lecithinase)

lysis of the
vacuole (LLO)

phagocytosis

intracellular
multiplication

intracellular
movement (ActA)

cell-cell spreading

Fig. 1. Successive steps of the *L. monocytogenes* infectious process. The proteins involved in this process are indicated in parentheses. *LLO*, listeriolysin O. (Adapted from TILNEY and PORTNOY 1989)

spread of the bacterium (for reviews see COSSART 1995; COSSART and KOCKS 1994; TILNEY and TILNEY 1993) and in which the organism never leaves the mammalian cytosol. This mechanism of cell-to-cell spreading was also shown to occur in nonphagocytic cells, using the human intestinal epithelial cell line Caco-2 (MOUNIER et al. 1990).

4 Entry into Nonphagocytic Cells

Based on both in vivo and tissue culture studies, it is clear that *L. mono-cytogenes* enters into nonphagocytic cells. Among the various cell lines which have been used to study the invasive process (for a review see COSSART and MENGAUD 1989; SHEEHAN et al. 1994) the human enterocyte Caco-2 cell line is a widely employed in vitro model, because it has the remarkable capacity to undergo typical enterocyte differentiation (e.g., polarization, formation of micro-villi and tight junctions) under standard culture conditions. A recent electron microscopic study showed that *L. monocytogenes* was able to enter Caco-2 cells through the apical pole (KARUNASAGAR et al. 1994). This study employed non-differentiated Caco-2 cells and is not in agreement with several other studies which suggested that *L. monocytogenes* enter Caco-2 (GAILLARD et al. 1994) and LLC-PK1 cells through the baso-lateral surface (TEMM-GROVE et al. 1994). These latter results are corroborated by the fact that confluent Caco-2 cells are not permissive to *L. monocytogenes* infection (our unpublished results).

L. *monocytogenes* is an adherent bacterium, and the invasion process may be artificially divided into two steps: attachment to the host cell surface and subsequent internalization. At 4°C, bacteria adhere but do not enter. Bacterial entry occurs when the temperature is raised (our unpublished results). Uncoupling of attachment and entry are also attained by treating cells with cytochalasin D, which inhibits entry but not adherence (GAILLARD et al. 1987). These results demonstrate that microfilament function is required only for uptake and suggest that bacterial entry occurs by a process closely related to phagocytosis.

For *L. monocytogenes*, nothing is known about bacterial factors which are involved only in adherence. To date, three surface proteins, internalin (the *inlA* gene product), InlB, and p60, have been reported to play a role in the induced internalization of *L. monocytogenes* by nonprofessional phagocytic cells in vitro.

4.1 The Internalin Gene Family

Discovery of the internalin gene family resulted directly from the identification of the first invasion locus in *L. monocytogenes*, the internalin gene (GAILLARD et al. 1991). A library of Tn*1545* mutants of *L. monocytogenes* was screened for defects in entry into Caco-2 cells. Three noninvasive mutants were obtained. These mutants were also defective for entry into a variety of other epithelial cell lines. In all three mutants, the transposon had inserted into a region upstream from two open reading frames, *inlA* and *inlB*. Transcription of these two genes was abolished in the noninvasive mutants. Complementation of the noninvasive mutants with *inlA* restored invasivity in Caco-2 cells. In addition, when *inlA* was expressed in *L. innocua* it conferred invasivity on this otherwise noninvasive, nonpathogenic *Listeria* species. Thus, at least in the genetic background of a closely related *Listeria* species, *inlA* expression is sufficient to promote entry into intestinal epithelial cells.

inlA encodes internalin, an 800-amino acid protein (DRAMSI et al. 1993a; GAILLARD et al. 1991) whose characteristic features include a signal sequence, two different regions of repeats, and a COOH-terminal hydrophobic region preceded by the hexapeptide LPTTGD (Fig. 2). The first region of repeats is made of 15 highly conserved successive repeats of 22 amino acids which display a striking periodicity of leucine residues. Such repeats have been termed "LRRs", for leu-rich repeats (KOBE and DEISENHOFER 1994). The second region of repeats is formed by three successive repeats, the first two of 70 amino acids and the third

→

Fig. 2A,B. The internalin locus. **A** the internalin operon and the putative terminators (*T1* and *T2*) are indicated. Also shown is the approximate location of the Tn*1545* insertion used to identify the locus. Below *inlA* and *inlB*, a schematic representation of their corresponding proteins is shown. **B** Sequences of internalin and InlB showing signal sequences (*underlined*), repeat regions, and LPXTGX motifs (*boxed*). (From GAILLARD 1991)

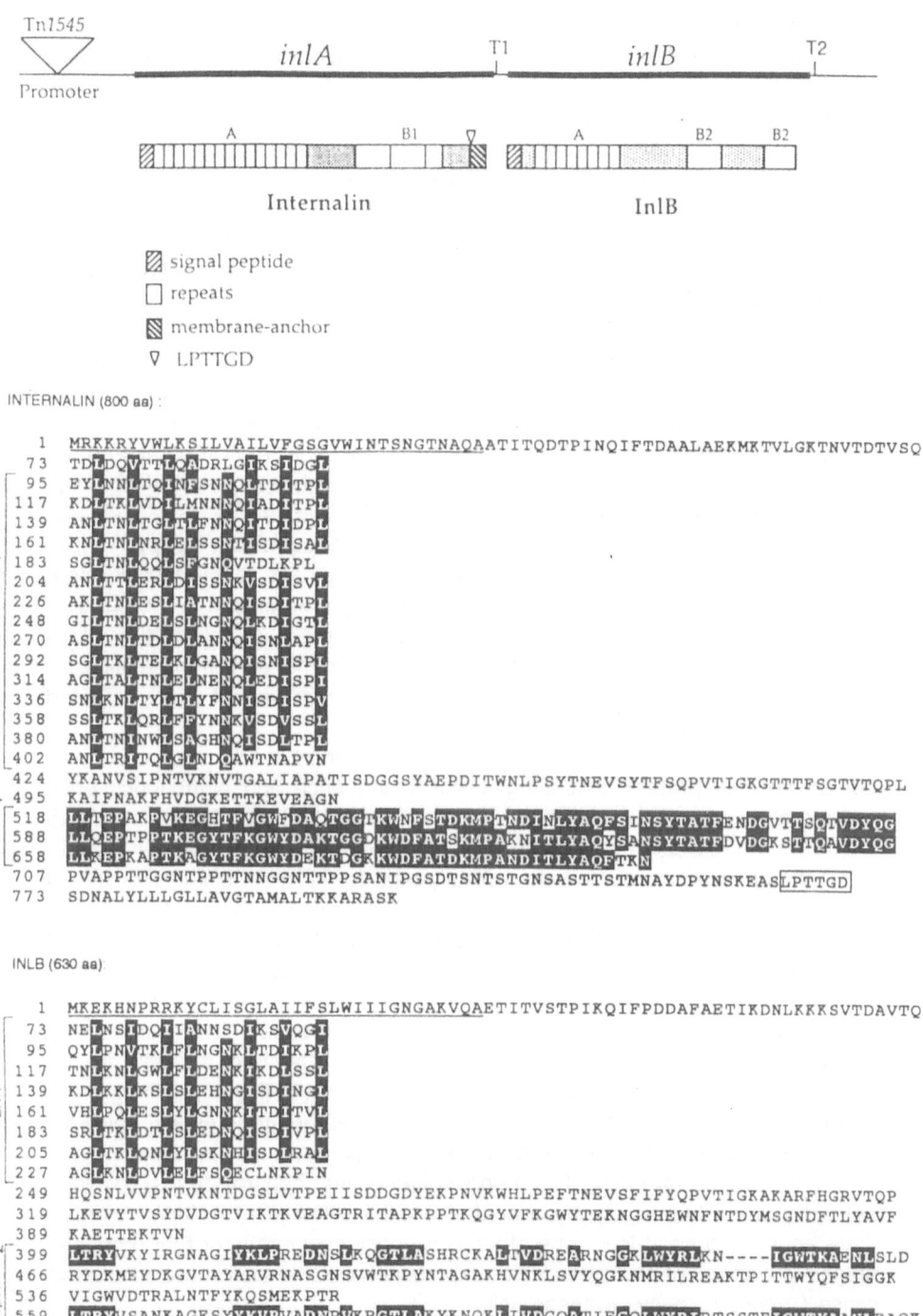

of 49 amino acids. The hexapeptide LPTTGD present in internalin fits a consensus hexapeptide LPXTGX (where X represents any amino acid) found in many surface proteins of gram-positive cocci [e.g., protein A of *Staphylococcus aureus* and protein M of *Streptococcus pyogenes* (FISCHETTI et al. 1990)]. Recently, NAVARRE and SCHNEEWIND demonstrated that this consensus sequence, together with an approximately 20-amino acid hydrophobic C-terminal region (also present in internalin), serves as a sorting signal to the cell wall (NAVARRE and SCHNEEWIND 1994).

Based on these studies, internalin is predicted to be a cell-wall-associated protein. Indeed, Western blot analysis has shown that internalin is a surface protein (DRAMSI et al. 1993b). These results were confirmed by immunogold labeling and immunofluorescence of in vitro grown bacteria using monoclonal antibodies against purified internalin (J. MENGAUD, F. NATO and P. COSSART our unpublished data). Maximal levels of the surface-bound form of internalin are found in exponential cultures, and this optimal expression of internalin on the bacterial surface in exponential phase correlates not only with maximal transcription, but also with maximal invasivity in Caco-2 cells (DRAMSI et al. 1993b) (Fig. 3). Surprisingly, internalin is also found in the culture supernatant, but preliminary results indicate that the released form does not play any role in invasion (our unpublished results). The mechanism of anchoring of internalin in the cell wall seems to be very similar to that of other LPXTGX-containing proteins (Fig. 4), as demonstrated by an experiment in which the C-terminal part of internalin was fused to that of protein A of *S.aureus*. The fused protein was efficiently targeted to the cell wall of *S. aureus* (SCHNEEWIND et al. 1993).

InlB is downstream from and co-transcribed with *inlA*. This gene encodes a 630-amino acid protein containing a signal sequence and eight leucine-rich repeats strikingly similar to those found in internalin (Fig. 2). However, unlike internalin, the InlB protein does not have a hydrophobic COOH-terminal region.

Although it is not located immediately next to the gene cluster containing most of the other identified *L. monocytogenes* virulence factors (MICHEL and COSSART 1992), the *inlAB* locus is co-regulated with these virulence genes and shares dependence on the pleiotropic virulence gene activator PrfA (DRAMSI et al. 1993b).

The role of *inlB* was assessed through the construction of isogenic mutants containing chromosomal deletions in the *inlAB* locus (DRAMSI et al. 1995). These studies showed that: (a) *inlB* encodes a surface protein (Fig. 5); (b) internalin, which was previously shown to be necessary for entry into intestinal epithelial cells, is also required for entry into the human hepatocyte cell line HepG-2; and (c) *inlB* is required for entry of *L. monocytogenes* into hepatocyte-like cell lines but not into the human intestinal epithelial cell line Caco-2. in addition to InlB, there appears to be at least one other *L. monocytogenes* factor that mediates entry into cultured hepatocytes, since expression of *inlB* alone is not sufficient to cause *L. innocua* to invade these cells.

To evaluate the role of internalin and InlB in vivo, BALB/c mice were inoculated intragastrically with either the wild-type strain EGD or the isogenic

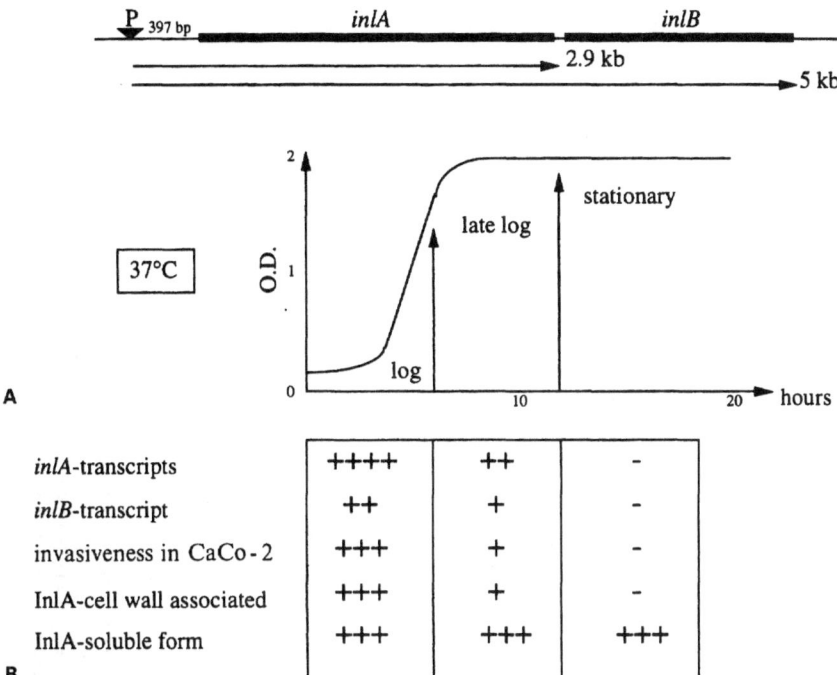

Fig. 3A,B. Expression of internalin. **A** Transcriptional organization of the internalin locus. The two transcripts of the internalin operon are indicated by *arrows*, and the promotor region is indicated as *P*. **B** Expression of the internalin locus and invasiveness of *L. monocytogenes* in Caco-2 cell line during growth

strain ΔinlAB. Inoculation with the mutant resulted in reduced bacterial counts in the liver compared with inoculation with the wild-type strain, demonstrating that the *inlAB* locus is important for normal infection of this organ. Bacterial counts in the mesenteric lymph nodes and the spleen, however, were very similar over a period of 7 days (DRAMSI et al. to be published). Thus the *inlAB* locus is apparently not essential for penetration of the epithelial cell barrier. Penetration of this barrier in the absence of *inlAB* could reflect the existence of genetic redundancy, reinforcing the view that this insidious pathogen uses multiple strategies to invade cells and tissues.

In addition to *inlB*, five other genes homologous to *inlA* have been identified, cloned, and sequenced (S. DRAMSI, P. DEHOUX, and P. COSSART, unpublished results). Our working hypothesis is that the internalin repertoire encodes surface proteins with different cellular and/or species tropisms. Our results with InlB favor such a hypothesis.

Internalin, InlB, and the other products of the internalin multigene family do not display striking overall homology to other known proteins. However, all but one member of this family contain the series of 22 amino acid LRRs described above for internalin and InlB. LRRs are present in a number of eukaryotic proteins

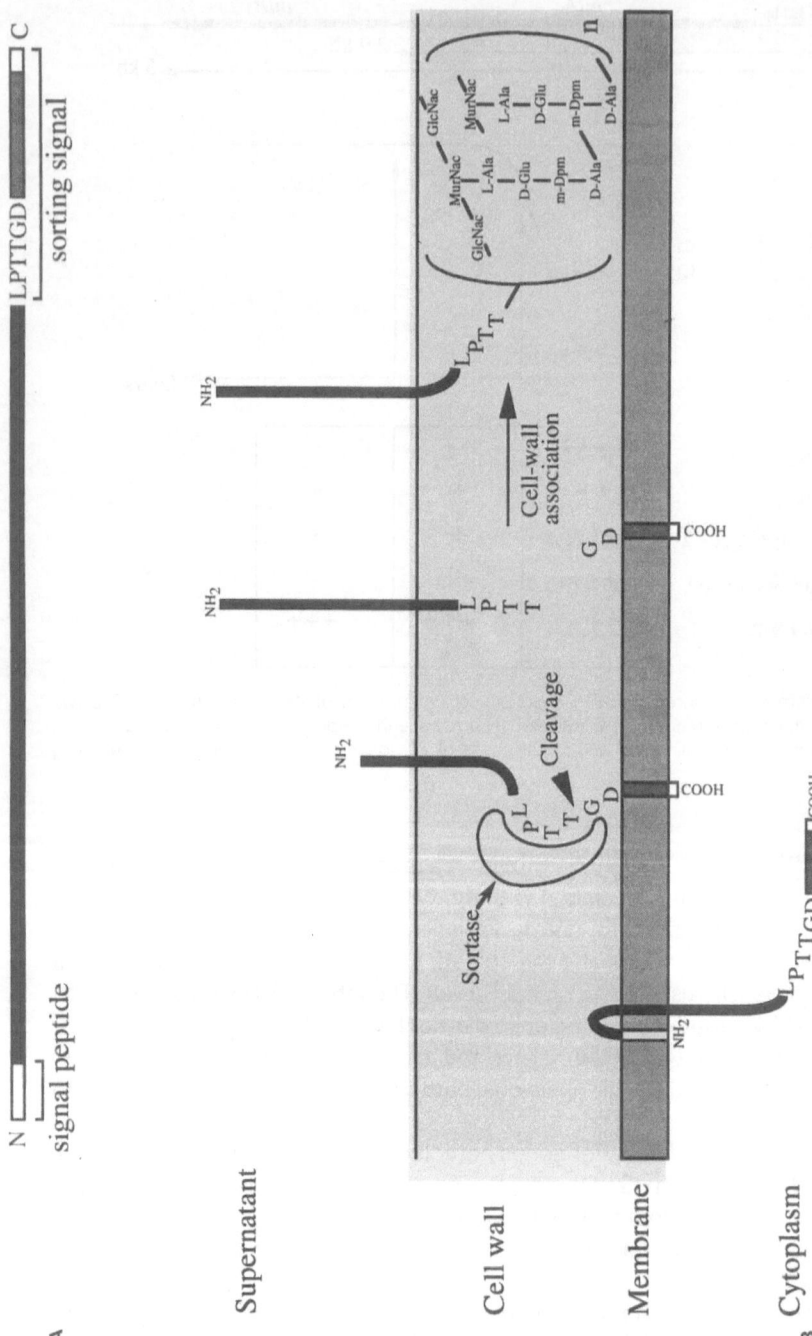

Fig. 4A,B. Anchoring of internalin in the cell wall of *L. monocytogenes*. **A** Schematic representation of internalin. **B** Putative model of the cell wall sorting of internalin (adapted from NAVARRE and SCHNEEWIND 1994). Internalin is exported to the surface of the bacterium. After cleavage of the peptide signal, internalin is retained in the membrane by a C-terminal domain consisting of a hydrophobic region and a charged tail. The LPTTGD motif is then recognized by an unidentified cell wall-sorting machinery, a "sortase" (NAVARRE and SCHNEEWIND 1994) which cleaves the polypeptide chain between threonine (*T*) and glycine (*G*) of the LPXTGX consensus sequence. The N-terminal fragment of the precursor protein is then linked to the peptidoglycan of the *L. monocytogenes* cell wall

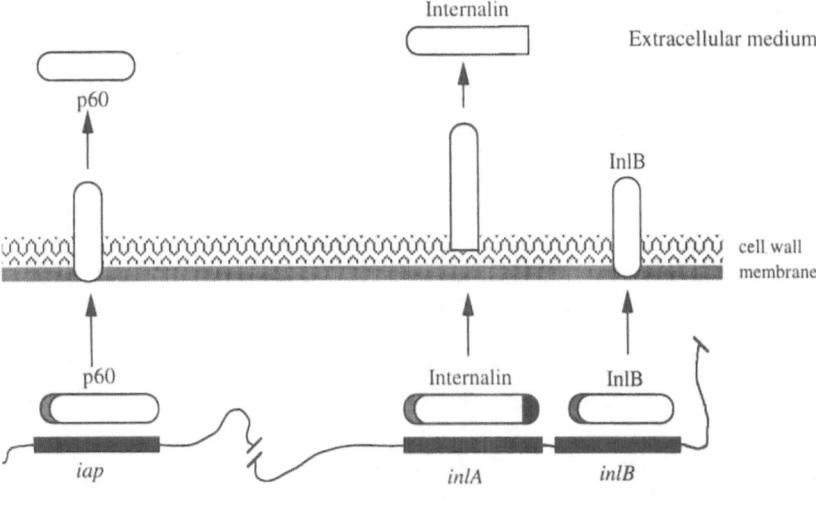

Fig. 5. The *iap* and *inlAB* genes of *L. monocytogenes* and location of their products

with diverse functions and cellular locations (KOBE and DEISENHOFER 1994). Apart from the products of the internalin gene family in *L. monocytogenes*, only two other examples of prokaryotic proteins containing LRRs have been described, the IpaH surface proteins of *S. flexneri* (HARTMAN et al. 1990) and the secreted YopM protein of *Yersinia pestis* (LEUNG and STRALEY 1989; VENKATESAN et al. 1991).

LRRs are usually present in tandem, and their number ranges from one in platelet glycoprotein Ib (LOPEZ et al. 1988) to 30 in chaoptin (REINKE et al. 1988). The most common length of an LRR is 24 residues, but repeats containing any number of residues between 20 and 29 have been identified. LRR-containing proteins are often involved in strong protein-protein interactions. Recently, the crystal structure of the LRR-containing protein porcine ribonuclease inhibitor was solved. This structure revealed that in this protein, which consists entirely of LRRs of 28–29 residues, each LRR forms a β sheet followed by an α helix (KOBE and DEISENHOFER 1993).

There is evidence that LRRs could adopt other conformations as well. In fact, predictive analysis has suggested (F. JURNAK personal communication) that the LRRs present in internalin could fold into a new type of helix observed for the first time in the structure of pectate lyase of *Erwinia chrysanthemi* (COHEN 1993; YODER et al. 1993). This novel helix (termed a β helix) is right handed and has 22 amino acids per turn with a pitch of 4.8°A and a rise of 0.22°A per residue. (For comparison, the right-handed classical α helix has 3.6 residues per turn, a pitch of 5.4°A, and a rise of 1.5°A per residue.) The role of the LRRs in the structure and function of internalin is currently under investigation.

4.2 Protein p60

p60 is a major extracellular protein, first described as an invasion-associated protein (KUHN and GOEBEL 1989) because strains affected in p60 production were defective for invasion. However, more recent evidence indicates that the role of p60 in the entry process is likely to be indirect.

p60 mutants form long chains in which the bacteria are separated by double septa. These chains disaggregate to normal-sized single bacteria upon treatment with partially purified p60, and these p60 disaggregated bacteria, but not those obtained by physical disruption (ultrasonication), are able to invade 3T6 cells (KUHN and GOEBEL 1989). Analysis of the gene coding for p60, *iap*, revealed that p60 is a protein of 484 amino acids with a putative signal sequence and an extended repeat region consisting of 19 threonine-asparagine units (KÖHLER et al. 1990, 1991). Sequences homologous to *iap* are present in the other *Listeria* species (BUBERT et al. 1992). Like internalin, p60 is found in the culture supernatant and associated with the cell wall (RUHLAND et al. 1993) (Fig. 2). Recently, it was shown that *iap* is essential and that its product, p60, has murein hydrolase activity (WUENSCHER et al. 1993). Taken together, these results indicate that *iap* plays a role in bacterial growth, possibly in normal septum formation or separation of daughter cells, and suggest that the effect of mutations in *iap* on invasion are indirect.

5 Concluding Remarks

To date, the cellular aspects of the mechanisms of entry of *L. monocytogenes* into non-phagocytic cells have not been extensively investigated. A recent study has shown that immortalized cell lines are more permissive than finite or primary cells, suggesting that the state of differentiation may influence expression of specific receptors, and thus entry (VELGE et al. 1994a). This observation deserves further investigation. Two reports indicate that tyrosine kinase inhibitors block entry of *L. monocytogenes* (TANG et al. 1994; VELGE et al. 1994b), and one of these further suggested that MAP kinase was phosphorylated upon attachment of *L. monocytogenes* to epithelial cells (TANG et al. 1994). However, this phosphorylation was later shown to be due solely to the action of listeriolysin O on the cell membrane (P. TANG and B.B. FINLAY personal communication). Thus far, no correlation between phosphorylation of a specific protein and the entry of *L. monocytogenes* has been established.

It is clear that rearrangement of the cytoskeleton is essential for entry of *L. monocytogenes* into normally nonphagocytic cells since, as previously mentioned, bacterial uptake into such cells is inhibited by treatment with cyto-chalasin D (GAILLARD et al. 1987). However, in contrast to the spectacular membrane ruffles detected in the case of *Salmonella* (FRANCIS et al. 1992, 1993) or *Shigella* (ADAM et al. 1995), preliminary observations indicate that no such structures are observed upon entry of *L. monocytogenes* into the host cell (J. MENGAUD, B. OHAYON, and P. COSSART unpublished results). Thus it is likely that invasion by *Listeria* occurs by mechanisms that are different from those employed by these gram-negative pathogens.

As detailed in this review, internalin is the best-characterized protein involved in entry of *L. monocytogenes* into nonphagocytic cells. Formerly, bacterial proteins involved in entry had been studied mainly in the case of *Shigella, Salmonella,* and *Yersinia* species (for reviews see (FALKOW et al. 1992; FINLAY 1994; MILLER and PEPE 1994; PARSOT 1994; PIERSON 1995). However, the only protein demonstrated to be sufficient to mediate entry of a bacterial pathogen into mammalian cells is invasin, a 986-amino acid outer membrane protein of *Yersinia psedotuberculosis* (ISBERG 1989, 1991). Internalin and invasin do not share similarity except for the presence of the pentapeptide FATDK, which is present in one copy at position 831–835 in invasin and in three copies (one perfect and two slightly degenerate) in the B repeats of internalin. Invasin binds to β1 integrin receptors (ISBERG 1989, 1991), and this interaction appears to be sufficient for bacterial entry. The receptor for internalin has not yet been identified.

In the cases of *Salmonella* (FINLAY 1994) and *Shigella* (PARSOT 1994), the entry process is clearly multifactorial. The genes required for invasion by *Shigella* are clustered in an approximately 30-kb region of the so-called virulence plasmid. This region contains 33 genes clustered in two regions, the *mxi-spa* region and the *ipa* region. The IpaB, IpaC, and IpaD proteins are prime candidates for proteins that mediate invasion. They were first shown to be associated with the outer membrane of the bacterium but are, in fact, also found in the culture medium. They do not have a signal sequence, and their secretion depends on the *mxi-spa* secretion locus. Thus, the *mxi* and *spa* mutants, like the *ipa* mutants, are not invasive due to their inability to secrete the Ipa proteins. Interestingly, sequence comparisons have revealed extensive similarities between some Mxi and Spa proteins and *Yersinia* proteins involved in the secretion of the Yop proteins which are encoded by the *Yersinia* virulence plasmid. In addition, representatives of the *Shigella* Mxi and Spa proteins are also present in S*almonella typhimurium*, and these homologues are also required for invasion by these bacteria. Recent results suggest that direct secretion of Ipas on or into the target cell might be the mechanism by which the pathogen induces the signals leading to cytoskeleton rearrangement and *Shigella* uptake. The number of genes involved and the existence of similar proteins suggest that the mechanisms of invasion by *Salmonella* and *Shigella* share similarities. Our present knowledge suggests that *L. monocytogenes* may use mechanisms that are quite different from those employed by these two gram-negative bacteria to gain entry into the host cell.

As a final point, it is important to indicate that the mechanism of cell-to-cell spreading, thus far described mostly in tissue culture models, could play a very important role in mediating bacterial spreading among different cell types in vivo. For example, it could well be that the presence of *L. monocytogenes* in intestinal cells after oral infection is due to direct cell-to-cell spreading from the M cells to the enterocytes, rather than to direct infection (invasion) of enterocytes. It is also possible that in the liver, entry into hepatocytes is due mainly to direct spreading from the Kupffer cells to the hepatocytes. Given (a) the multiple mechanisms of cellular infection (e.g., invasion mediated by members of the internalin gene family, direct intracytosolic cell-to-cell spreading) readily available to this pathogen, (b) the large number of tissues that are infected (cell and tissue tropism), and (c) the wide host range of this bacterium (species tropism), the biology of *L. monocytogenes* infection and dissemination in the infected animal is likely to be highly complex.

Acknowledgments. We acknowledge K. Ireton, J. Mengaud, and L. Braun for their critical reading of the manuscript. We thank F. Jurnak for personal communication of the analysis of the structure of internalin. Our work received support from DRET contrat 93–109, INSERM CRE 93013, CEE SC1-CT91-0682, CNRS URA 1300 and the Pasteur Institute.

References

Adam T, Arpin M, Prévost MC, Gounon P, Sansonetti PJ (1995) Cytoskeletal rearrangements during entry of *Shigella flexneri* into HeLa cells. J Cell Biol (in press)

Alford CA, King TE, Campbell PA (1991) Role of transferrin, transferrin receptor and iron in macrophage listericidal activity. J Exp Med 174: 459–456

Alvarez-Dominguez E, Carasco-Marin E, Leyva-Cobian F (1993) Role of complement component C1q in phagocytosis of *Listeria monocytogenes* by murine macrophage-like cell lines. Infect Immun 61: 3664–3672

Audurier A, Pardon P, Marly J, Lantier F (1980) Experimental infection of mice with *Listeria monocytogenes* and *L. innocua*. Ann Microbiol 131B: 47–57

Baker LA, Campbell PA (1980) Thioglycolate medium decreases resistence to bacterial infection in mice. Infect Immun 27: 455–460

Beckerman KP, Rogers HW, Corbett JA, Schreiber RD, McDaniel ML, Unanue ER (1993) Release of nitric oxide during the T cell-independent pathway of macrophage activation. J Immunol 150: 888–895

Bliska JB, Galan JE, Falkow S (1993) Signal transduction in mammalian cell during bacterial attachment and entry. Cell 73: 903–920

Boockvar KS, Granger DL, Poston RM, Maybodi M, Washington MK, Hibbs JBJ, Kurlander RL (1994) Nitric oxide produced during murine listeriosis is protective. Infect Immun 62: 1089–1100

Bubert A, Kuhn M, Goebel W, Kohler S (1992) Structural and functional properties of the p60 proteins from different *Listeria* species. J Bacteriol 174: 8166–8171

Campbell PA (1994) Macrophage-*Listeria* interactions. In: Zwilling BS, Eisenstein TK (eds) Macrophage-pathogen interactions. Dekker, Newyork pp 313–328 (Immunology, vol 60)

Campbell PA, Czuprynski CJ, Cook JL (1984) Differential expression of macrophage functions: bactericidal versus tumoricidal activities. J Leukoc Biol 36: 293–306

Carter PS, Collins FM (1974) The route of enteric infection in normal mice. J Exp Med 139: 1189–1203

Cohen FE (1993) The parallel β helix of pectate lyase C: something to sneeze at. Science 260: 1444–1445

Conlan JW, North RJ (1991) Neutrophil-mediated dissolution of infected host cells as a defense strategy against a facultative intracellular bacterium. J Exp Med 174: 741–744

Cossart P (1994) *Listeria monocytogenes*: strategies for entry and survival in cells and tissues. In: Russel DG (ed) Clinical infectious diseases: strategies for intracellular survival of microbes, vol. 1 Baillére Tindall, London, pp 285–304

Cossart P (1995) Bacterial actin-based motility. Curr Opin Cell Biol 7: 94–101

Cossart P, Kocks C (1994) The actin-based motility of the intracellular pathogen *Listeria monocytogenes*. Mol Microbiol 13: 395–402

Cossart P, Mengaud J (1989) *Listeria monocytogenes*: a model system for the molecular study of intracellular parasitism. Mol Biol Med 6: 463–474

Croizè J, Arvieux J. Berche P, Colomb MG (1993) Activation of the human complement alternative pathway by *Listeria monocytogenes*: evidence for direct binding and proteolysis of the C3 component on bacteria. Infect Immun 61: 5134–5139

De Chastellier C, Berche P (1994) Fate of *Listeria monocytogenes* in murine macrophages: evidence for simultaneous killing and survival of intracellular bacteria. Infect Immun 62: 543–553

Dramsi S, Dehoux P, Cossart P (1993a) Common features of gram-positive bacterial proteins involved in cell recognition. Mol Microbiol 9: 1119–1122

Dramsi S, Kocks C, Forestier C, Cossart P (1993b) Internalin-mediated invasion of epithelial cells by *Listeria monocytogenes* is regulated by the bacterial growth state, temperature and the pleiotropic activator, PrfA. Mol Microbiol 9: 931–941

Dramsi S, Biswas I, Maguin E, Mastroeni P, Cossart P (1995) Entry of *Listeria monocytogenes* into hepatocytes requires expression of InlB, a surface protein of the internalin multigene family. Mol Microbiol (in press)

Drevets D, Campbell PA (1991) Roles of complement and complement receptor type 3 in phagocytosis of *Listeria monocytogenes* by inflammatory mouse macrophages. Infect Immun 59: 2645–2652

Drevets D, Canono BP, Campbell PA (1992) Listericidal and nonlistericidal mouse macrophages differ in complement receptor type 3-mediated phagocytosis of *L. monocytogenes* and in preventing escape of the bacteria into the cytoplasm. J Leukoc Biol 52: 70–79

Drevets DA, Leenen PJM, Campbell PA (1993) Complement receptor type 3(CD11b/CD18) involvement is essential for killing of *Listeria monocytogenes* by mouse macrophages. J Immunol 151: 5431–5439

Dunne DW, Resnik D, Greenberg J, Krieger M, Joine K (1994) The type I macrophage scavenger receptor binds to gram-positive bacteria and recognizes lipotechoic acids. Proc Natl Acad Sci USA 91: 1863–1867

Falkow S, Isberg RR, Portnoy DA (1992) The interaction of bacteria with mammalian cells. Annu Rev Cell Biol 8: 333–363

Finlay BB (1994) Molecular and cellular mechanism of *Salmonella* pathogenesis. Curr Top Microbiol Immunol 192: 163–185

Fischetti VA, Pancholi V, Schneewind O (1990) Conservation of a hexapeptide sequence in the anchor region of surface proteins from gram-positive cocci. Mol Microbiol 4: 1603–1605

Fleming DW, Cochi SL, MacDonald KL, Brondum J, Hayes PS, Plikaytis BD, Holmes MB, Audurier A, Broome CV, Reingold AL (1985) Pasteurized milk as a vehicle of infection in an outbreak of listeriosis. N Engl J Med 312: 404–407

Francis CL, Starnbach MN, Falkow S (1992) Morphological and cytoskeletal changes in epithelial cells occur immediately upon interaction with *Salmonella typhimurium* grown under low-oxygen conditions. Mol Microbiol 6: 3077–3087

Francis CL, Ryan TA, Jones BD, Smith SJ, Falkow S (1993) Ruffles induced by *Salmonella* and other stimuli direct macropinocytosis of bacteria. Nature 364: 639–642

Gaillard JL, Berche P, Mounier J, Richard S, Sansonetti PJ (1987) *In vitro* model of penetration and intracellular growth of *L. monocytogenes* in the human enterocyte-like cell line Caco-2. Infect Immun 55: 2822–2829

Gaillard J-L, Berche P, Frehel C, Gouin E, Cossart P (1991) Entry of *L. monocytogenes* into cells is mediated by internalin, a repeat protein reminiscent of surface antigens from gram-positive cocci. Cell 65: 1127–1141

Gaillard JL, Dramsi S, Berche P, Cossart P (1994) Molecular cloning and expression of internalin in *Listeria*. Methods Enzymol 236: 551–565

Gray ML, Killinger AH (1966) *Listeria monocytogenes* and listeric infections. Bacteriol Rev 30: 309–382

Grutzkau A, Hanski C, Hahn H, Riecken EO (1990) Involvement of M cells in the bacterial invasion of Peyer's patches: a common mechanism shared by *Yersinia enterocolitica* and other invasive bacteria. Gut 31: 1011–1015

Hanski C, Kutschka U, Schmoranzer HP, Naumann M, Stallmach A, Hahn H, Menge H, Riecken EO (1989) Immunohistochemical and electron microscopic study of interactions of *Yersinia enterocolitica* serotype O8 with intestinal mucosa during experimental enteritis. Infect Immun 57: 673–678

Hartman AB, Venkatesan M, Oaks EV, Buysse JM (1990) Sequence and molecular characterization of a multicopy invasion plasmid antigen gene, *ipaH*, of *Shigella flexneri*. J Bacteriol 172: 1905–1915

Isberg RR (1989) Mammalian cell adhesion functions and cellular penetration of enteropathogenic *Yersinia* species. Mol Microbiol 3: 1449–1453

Isberg RR (1991) Discrimination between intracellular uptake and surface adhesion of bacterial pathogens. Science 252: 934–938

Karunasagar I, Senghaas B, Krohne G, Goebel W (1994) Ultrastructural study of *Listeria monocytogenes* entry into cultured human colonic epithelial cells. Infect Immun 62: 3554–3558

Kaufmann SHE (1993) Immunity to intracellular bacteria. Annu Rev Immunol 11: 129–163

Kobe B, Deisenhofer J (1993) Crystal structure of porcine ribonuclease inhibitor, a protein with leucine-rich repeats. Nature 366: 751–756

Kobe B, Deisenhofer J (1994) The leucine-rich repeat: a versatile binding motif. TIBS 19: 415–420

Köhler S, Leimeister-Wächter M, Chakraborty T, Lottspeich F, Goebel W (1990) The gene coding for protein p60 of *Listeria monocytogenes* and its use as a specific probe for *Listeria monocytogenes*. Infect Immun 58: 1943–1950

Köhler S, Bubert A, Vogel M, Goebel W (1991) Expression of the iap gene coding for protein p60 of *Listeria monocytogenes* is controlled on the posttranscriptional level. J Bacteriol 173: 4668–4674

Kuhn M, Goebel W (1989) Identification of an extracellular protein of *Listeria monocytogenes* possibly involved in intracellular uptake by mammalian cells. Infect Immun 57: 55–61

Kuhn M, Kathariou S, Goebel W (1988) Hemolysin supports survival but not entry of the intracellular bacterium *Listeria monocytogenes*. Infect Immun 56: 79–82

Lepay DA, Steinman RM, Nathan CF, Murray HW, Cohn ZA (1985) Liver macrophages in murine infection. Cell-mediated immunity is correlated with an influx of macrophages capable of generating reactive oxygen intermidiates. J Exp Med 161: 1503–1512

Leung KY, Straley SC (1989) The yopM gene of *Yersinia pestis* encodes a released protein having homology with the human platelet surface protein GPIba. J Bacteriol 171: 4623–4632

Lopez JA, Chung DW, Fujikawa K, Hagen FS, Davie EW, Roth G (1988) The α and β chains of human platelet glycoprotein Ib are both transmembrane proteins containing a leucine-rich amino acid sequence. Proc Natl Acad Sci USA 85: 2135–2139

Mac Donald TT, Carter PB (1980) Cell-mediated immunity to intestinal infection. Infect Immun 28: 516–523

Mackaness GB (1962) Cellular resistance to infection. J Exp Med 116: 381–406

Mackaness GB (1964) The immunological basis of acquired cellular resistance. J Exp Med 120: 105–120

Macro AJ, Prats N, Ramos JA, Briones V, Blanco M, Dominguez L, Domingo M (1992) A microbiological, histopathological and immunohistological study of the intragastric inoculation of *L. monocytogenes* in mice. J Comp Pathol 107: 1–9

Miake S, Takeya K, Matsumoto T, Yoshikai Y, Nomoto K (1980) Relation between bactericidal and phagocytic activities of peritoneal macrophages induced by irritants. J Reticuloendothel Soc 27: 421–427

Michel E, Cossart P (1992) Physical map of the *Listeria monocytogenes* chromosome. J Bacteriol 174: 7098–7103

Miller VL, Pepe JC (1994) The invasion genes of *Yersinia*: inv, ail and yadA. In: Russel DG (ed) Clinical infectious diseases: strategies for intracellular survival of microbes, vol I. Baillére Tindall, London, pp 285–304

Mounier J, Ryter A, Coquis-Rondon M, Sansonetti PJ (1990) Intracellular and cell-to-cell spread *Listeria monocytogenes* involves interaction with F-actin in the enterocyte-like cell line Caco-2. Infect Immun 58: 1048–1058

Navarre WW, Schneewind O (1994) Proteolytic cleavage and cell wall anchoring at the LPXTGX motif of surface proteins in gram-positive bacteria. Mol Microbiol 14: 115–121

North RJ (1970) The relative importance of blood monocytes and fixed macrophages to the expression of cell-mediated immunity to infection. J Exp Med 132: 521–534

North RJ (1973) Cellular mediators of anti-*Listeria* immunity as an enlarged population of short-lived, replicating T cells. Kinetics of their production. J Exp Med 138: 342–355

Parsot C (1994) *Shigella flexneri*: genetics of entry and intracellular dissemination in epithelial cells. Top Microbiol Immunol 192: 217–241

Pierson DE (1995) Mechanisms of *Yersinia* entry into mammalian cells. In: Miller VL, Kaper JB, Portnoy DA, Isberg RR (eds) Molecular genetics of bacterial pathogenesis. American Society of Microbiology, Washington DC, pp 235–247

Portnoy, DA (1995) Cellular biology of *Listeria monocytogenes* infection. In: Miller VL, Kaper JB, Portnoy DA, Isberg RR (eds) Molecular genetics of bacterial pathogenesis. American Society of Microbiology, Washington DC, pp 235–247

Portnoy DA, Jacks PS, Hinrichs D (1988) Role of hemolysin for the intracellular growth of *L. monocytogenes*. J Exp Med 167: 1459–1471

Portnoy DA, Schreiber RD, Connelly P, Tilney L (1989) Gamma interferon limits access of *L. monocytogenes* to the macrophage cytoplasm. J Exp Med 170: 2141–2146

Racz P, Tenner K, Szivessy K (1970) Electron microscopic studies in experimental keratoconjunctivitis listeriosa. I. Penetration of *Listeria monocytogenes* into corneal epithelial cells. Acta Microbiol Acad Sci Hung 17: 221–236

Racz P, Tenner K, Mérö E (1972) Experimental *Listeria* enteritis. I. An electron microscopic study of the epithelial phase in experimental *Listeria* infection. Lab Invest 26: 694–700

Reinke R, Krantz DE, Yen D (1988) Chaoptin, a cell surface glycoprotein required for *Drosophila* photoreceptor cell morphogenesis, contains a repeat motif found in yeast and humans. Cell 52: 291–301

Rogers HW, Unanue ER (1993) Neutrophils are involved in acute nonspecific resistance to *Listeria monocytogenes* in mice. Infect Immun 61: 5090–5096

Rosen H, Gordon S, North RJ (1989) Exacerbation of murine listeriosis by a monoclonal antibody specific for the type 3 complement receptor of myelomonocytic cells. Absence of monocytes at infective foci allows *Listeria* to multiply in nonphagocytic cells. J Exp Med 170: 27–37

Ruhland GJ, Hellwig M, Wanner G, Fiedler F (1993) Cell-surface location of *Listeria*-specific protein p60—detection of *Listeria* cells by indirect immunofluorescence. J Gen Microbiol 139: 609–616

Schlech III WF, Lavigne PM, Bortolussi RA, Allen AC, Haldane VE, Wort JA, Hightower AW, Johnson SE, King SH, Nicholls ES, Broome C (1983) Epidemic listeriosis—evidence for transmission by food. N Engl J Med 308: 203–206

Schneewind O, Mihaylova-Petkov D, Model P (1993) Cell wall sorting signals in surface proteins of gram-positive bacteria. EMBO J 12: 4803–4811

Sheehan B, Kocks C, Dramsi S, Gouin E, Klarsfeld A, Mengaud J, Cossart P (1994) Molecular and genetic determinants of the *Listeria monocytogenes* infectious process. Curr Top Microbiol Immunol 192: 187–216

Spitalny GL (1981) Dissociation of bactericidal activity from other functions of activated macrophages in exudates induced by thioglycolate medium. Infect Immun 27: 455–460

Tang P, Rosenshine I, Finlay BB (1994) *Listeria monocytogenes*, an invasive bacterium, stimulates MAP kinase upon attachment to epithelials cells. Mol Biol Cell 5: 455–464

Temm-Grove CT, Jockusch B, Rohde M, Niebuhr K, Chakraborty T, Wehland J (1994) Exploitation of microfilament proteins by *Listeria monocytogenes*: microvillus-like composition of the comet tails and vectorial spreading in polarized epithelial sheets. J Cell Sci 107: 2951–2960

Tilney LG, Portnoy DA (1989) Actin filaments and the growth, movement, and spread of the intracellular bacterial parasite, *Listeria monocytogenes*. J Cell Biol 109: 1597–1608

Tilney LG, Tilney MS (1993) The wily ways of a parasite: induction of actin assembly by *Listeria*. Trends Microbiol 1: 25–31

Velge P, Bottreau E, Kaeffer B, Pardon P (1994a) Cell immortalization enhances *Listeria monocytogenes* invasion. Med Microbiol Immunol 183: 145–158

Velge P, Bottreau E, Kaeffer B, Yurdusev N, Pardon P, Van Langendonck N (1994b) Protein tyrosine kinase inhibitors block the entries of *Listeria monocytogenes* and *Listeria ivanovii* into epithelial cells. Microb Pathog 17: 37–50

Venkatesan MM, Buysse JM, Hartman AB (1991) Sequence variation in two ipaH genes of *Shigella flexneri* 5 and homology to the LRG-like family of proteins. Mol Microbiol 5: 2435–2445

Wassef JS, Keren DF, Mailloux JL (1989) Role of M cells in initial antigen uptake and in ulcer formation in the rabbit intestinal loop model of shigellosis. Infect Immun 57: 858–863

Wood S, Maroushek N, Czuprynski C (1993) Multiplication of *L. monocytogenes* in a murine hepatocyte cell line. Infect Immun 61: 3068–3072

Wright SD, Silverstein SC (1983) Receptors for C3b and C3bi promote phagocytosis but not the release of toxic oxygen from human phagocytes. J Exp Med 158: 2016–2023

Wuenscher M, Kohler S, Bubert A, Gerike U, Goebel W (1993) The iap gene of *Listeria monocytogenes* is essential for cell viability and its gene product, p60, has bacteriolytic activity. J Bacteriol 175: 3491–3501

Yoder MD, Keen NT, Jurnak F (1993) New domain motif: the structure of pectate lyase C, a secreted plant virulence factor. Science 260: 1503–1507

Entry of Enteropathogenic *Escherichia coli* into Host Cells

M.S. Donnenberg

1 Introduction

1.1 Background

The enteropathogenic *Escherichia coli* (EPEC) are a class of diarrheagenic *E. coli* strains that induce a characteristic "attaching and effacing" effect on the host cell cytoskeleton and membrane, but do not produce high levels of Shiga-like toxins. EPEC strains are also distinguished by the "localized adherence" pattern by which they form discrete microcolonies on the surface of tissue culture cells. EPEC strains

Division of Infectious Diseases, University of Maryland School of Medicine, MSTF 900, 10 S. Pine Street, Baltimore, Maryland 21201; Medical Service, Veterans Affairs Medical Center, Baltimore, USA

are a leading cause of diarrhea worldwide, affecting primarily infants in developing countries (Cobeljić et al. 1989; Cravioto et al. 1991; Kain et al. 1991; Gomes et al. 1991; Echeverria et al. 1991; Levine et al. 1988). In some countries EPEC are the number one cause of infantile diarrhea, exceeding even rotavirus in incidence (Gomes et al. 1991; Robins-Browne et al. 1980; Cravioto et al. 1988). EPEC diarrhea is typically watery, often with associated vomiting and low-grade fever (Gomes et al. 1991). Diarrhea due to EPEC may be severe and protracted and may require hospitalization, intravenous rehydration, and total parental nutrition (Fagundes Neto et al. 1989; Hill et al. 1991). EPEC are thought to be transmitted primarily from person to person (Wu and Peng, 1992), but common-source food and water outbreaks have been described (Viljanen et al. 1990; Schtoeder et al. 1968). In developed countries outbreaks in child daycare centers have occurred (Paulozzi et al. 1986; Bower et al. 1989). Mortality due to EPEC infection continues to be a problem in developing countries (Senerwa et al. 1989).

1.2 Are EPEC Invasive?

EPEC are not invasive pathogens in the traditional sense. Diarrhea due to EPEC is not usually accompanied by blood or mucus in the stool. Fecal leukocytes are present in a minority of cases (Donnenberg et al. 1993a). Patients with EPEC rarely have bacteremia. Dissemination to extra intestinal sites has been described but is a near-terminal event in fatal cases (Giles et al. 1949; Bower et al. 1989; Wu and Peng 1992). Yet EPEC are capable of directing their internalization into host cells. This capacity for cellular invasion is easily demonstrable in vitro, where internalization efficiency can approach that of traditional invasive pathogens such as *Yersinia, Salmonella,* and *Shigella*[1]. In addition, intracellular bacteria have been observed by electron microscopy in tissue culture (Donnenberg et al. 1989; Miliotis et al. 1989; Andrade et al. 1989), in intestinal specimens from animals experimentally infected with EPEC (Tzipori et al. 1989; Moon et al. 1983; Staley et al. 1969; Polotsky et al. 1977), and even in biopsies from children with severe natural infection (Ulshen and Rollo 1980; Hill et al. 1991).

The capacity for epithelial cell invasion by EPEC is overshadowed by the ability of the organism to adhere to cells in great numbers. In comparison to enteroinvasive *E. coli*, a much smaller fraction of EPEC bacteria associated with tissue culture cells are internalized (Robins-Browne and Bennett-Wood 1992). However, even a small percentage of the large number of adherent bacteria results in a substantial total intracellular population.

One widely used EPEC strain, E2348/69, yields variable results in gentamicin invasion assays, depending upon the source of the strain. When acquired directly from the Central Public Health Laboratories, London, where it has been stored lyophilized since its isolation in 1969, the strain is resistant to streptomycin and spectinomycin, sensitive to nalidixic acid, and invades HEp-2 cells at approximately 5–10% of the inoculum. A version of the strain acquired from the Center for Vaccine Development, which had been stored on a slant for many years, is sensitive to streptomycin and spectinomycin, resistant to nalidixic acid, and invades HEp-2 cells at less than 1% of the inoculum. The latter strain, however, retains pathogenicity in volunteer studies

Perhaps the most convincing argument for inclusion of a chapter about EPEC in a volume devoted to bacterial invasiveness is the large amount of data relevant to bacterial pathogenesis that has been gleaned from the study of EPEC internalization. Since EPEC invasion can be dissected into discrete stages, the phenomenon has yielded to detailed genetic analysis and is now proving amenable to biochemical analysis as well. Thus, a tissue culture model of EPEC infection has become a paradigm for the study of bacterial interactions with epithelial cells including initial adherence, intimate attachment, signal transduction events, and, ultimately, internalization.

1.3 General Features of EPEC Invasion

EPEC are capable of invading a large number of cell types including poorly differentiated transformed human cell lines such as HeLa, Henle 407, and HEp-2, well-differentiated transformed human intestinal cells such as Caco-2 and T84, primary human kidney epithelial cells, and cells of animal origin such as Madin-Darby canine kidney cells and Chinese hamster ovary cells (DONNENBERG et al. 1989; MILIOTIS et al. 1989; ANDRADE et al. 1989; RILEY et al. 1990; FRANCIS et al. 1991; CANIL et al. 1993; author's unpublished data). As is the case for most bacteria capable of cell entry, internalization of EPEC can be inhibited by cytochalasins, which inhibit growth of host cell actin microfilaments (DONNENBERG et al. 1990b; ANDRADE et al. 1989; FRANCIS et al. 1991). Unlike many traditional invasive pathogens, however, the ability of EPEC to invade cells is also sensitive to inhibitors to microtubule function, suggesting that these structures are also involved in the entry process (DONNENBERG et al. 1990b; FRANCIS et al. 1991).

In contrast to enteroinvasive *E. coli*, EPEC are capable of entering epithelial cells at temperatures ranging from 32°C to 40°C (DONNENBERG et al. 1990b). Once inside epithelial cells, EPEC appear to remain inside membrane-bound vacuoles. The bacteria replicate intracellularly very slowly, if at all (DONNENBERG et al. 1990b; FRANCIS et al. 1991). EPEC are capable of passing through a polarized epithelial cell monolayer, but only after a prolonged lag period (CANIL et al. 1993). The inability of EPEC either to proliferate within cells or to pass rapidly through a monolayer may be interpreted as in vitro evidence in support of the notion that EPEC is primarily an extracellular pathogen.

1.4 A Collection of Noninvasive Mutants of EPEC

The study of EPEC pathogenesis has been greatly facilitated by the isolation of mutants deficient in interactions with epithelial cells (DONNENBERG et al. 1990a; JERSE et al. 1990). Among a collection of 322 TnphoA mutants of a virulent invasive EPEC strain, 22 are severely deficient for cellular invasion (DONNENBERG et al. 1990a). Each of these mutants produces an active alkaline phosphatase fusion protein, indicating that in each case the transposon has inserted into a gene that encodes an extra-cytoplasmic protein. However, five of the mutants have more than one transposon insertion; in each case, only one insertion per mutant

Table 1. Categories of noninvasive mutants of EPEC strain E2348/69

Category	Number of mutants	Defect	Known genes affected
1	7	Localized adherence	*bfpA*
2	4	Localized adherence	*dsbA*
3	7	Intimate attachment	*eaeA*
4	2	Signal transduction	*sepA*
5	2	?	?

results in an active fusion protein. Analysis of this collection by assays for localized adherence and attaching and effacing ability, as well as by subcloning the transposon fusion junctions, has led to the description of five categories (Table 1). From further study of these mutants, a model of EPEC pathogenesis has emerged that incorporates distinct stages of the interaction between bacterium and host. Throughout this chapter, information gained through study of these mutants will be highlighted as it relates to each step in pathogenesis.

2 Stages of Internalization

The interactions between EPEC and epithelial cells may be divided conceptually into three stages (Fig. 1). The first interaction to be considered results in the aggregation of bacteria into microcolonies on the surface of the cells in a pattern called localized adherence. The second stage consists of the signal transduction events that result in cytoskeletal damage. The third stage involves the more intimate attachment of the bacteria to the cells. The latter two steps are collectively known as the attaching and facing effect. Although it is tempting to consider these stages within the context of a temporal progression, in reality, the order in which they occur is not known. It may well be that the various stages take place simultaneously. To emphasize the rationale for distinguishing three stages, in the sections that follow intimate attachment will be discussed after initial adherence and before signal transduction. Finally, a subset of the bacteria that have engaged in these interactions is internalized into cells. It seems clear that invasion occurs after the three stages listed above are completed, since mutants deficient in any of these stages are defective for internalization and mutants capable of all three stages, but deficient in internalization, have been described.

2.1 Initial Adherence

CRAVIOTO et al. (1979) were the first to note that EPEC are capable of adhering avidly to tissue culture cells. Since the bacteria tend to form discrete microcolonies rather than blanketing the surface of the cells, the pattern displayed by EPEC has been termed "localized adherence" (SCALETSKY et al. 1984). Localized adherence is dependent upon the presence of a large plasmid highly conserved among EPEC strains of diverse clonal lineage (NATARO et al. 1987), which suggests

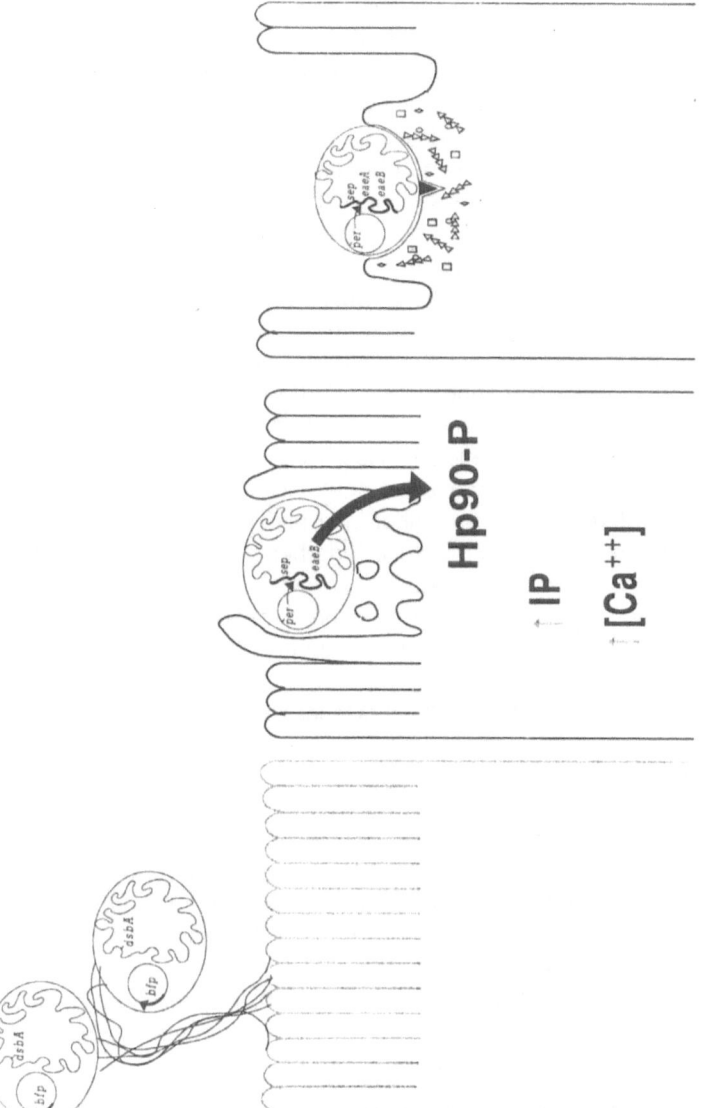

Fig. 1. A three-stage model of EPEC infection. Stages are presented as though representing a temporal progression but probably occur simultaneously. In the *first panel* localized adherence is mediated by bundle-forming pili encoded by the *bfp* gene cluster and additional plasmid loci but requires the chromosomal *dsbA* gene as well. In the *second panel*, soluble factors including the product of the *eaeB gene* are secreted by the products of the *seps* and initiate signal transduction events that include tyrosine phosphorylation of host cell proteins (*Hp90-p*), increased intracellular inositol phosphate (*IP*) and calcium (*Ca⁺⁺*) levels, and effacement of microvilli. Simultaneously, the *eae* gene cluster is activated by the product of the plasmid *per* locus. The third stage of infection (*third panel*) results when intimin (*solid triangle*), the product of the *eaeA* locus, mediates close attachment to the epithelial cell. From this proximity, effects on the epithelial cell are amplified with accumulation of filamentous actin and other cytoskeletal proteins (*geometric shapes*). (Adapted with permission from DONNENBERG and KAPER 1992)

that this plasmid has been acquired relatively recently by distantly related strains. Loss of the plasmid is associated with loss of localized adherence ability, while transfer of the plasmid to non-adherent laboratory *E. coli* strains confers this pheno- type (BALDINI et al. 1983). However, plasmid-cured EPEC strains remain capable of attaching and effacing activity, albeit with delayed kinetics and lowered efficiency in comparison to plasmid-containing strains (KNUTTON et al. 1987). NATARO et al. demonstrated that two large contiguous regions of the EPEC adherence plasmid are necessary for localized adherence. Recombinant strains containing both regions on separate plasmids are capable of localized adherence, whereas strains containing each region alone remain non-adherent (NATARO et al. 1987).

EPEC strains that possess the EPEC adherence plasmid are invasive for tissue culture cells, while strains that lack the plasmid, whether laboratory-cured or natural plasmid-free isolates, are approximately 100-fold less efficient at cell invasion (DONNENBERG et al. 1989). Among the collection of non invasive EPEC Tn*phoA* mutants are two categories deficient in localized adherence (DONNENBERG et al. 1990a). One group consists of seven mutants with insertions that map within 500 base pairs of the 90 000 bp adherence plasmid. The other group includes four mutants with insertions that map together on the EPEC chromo- some. Mutants from both groups, while deficient in initial adherence, neverthe- less remain capable of the subsequent steps of signal transduction and intimate attachment. The invasion defect in these mutants appears to result from ineffi- cient initial adherence which reduces the number of bacteria available for later events in pathogenesis. These mutants will be discussed in more detail below.

2.1.1 The Bundle-forming Pilus

Under specific conditions EPEC are capable of producing a flexible fimbria, termed the bundle-forming pilus (BFP) because of its tendency to form rope-like aggregates (GIRÓN et al. 1991). BFP are detected exclusively in strains of EPEC that possess the adherence plasmid. An antiserum raised against purified BFP partially inhibits localized adherence.

Analysis of the first group of invasion-deficient mutants resulted in the identification of the *bfpA* gene encoding bundlin, the major structural subunit of BFP (DONNENBERG et al. 1992). The *bfpA* gene is located near one end of one of the two fragments reported by NATARO et al. (1987) to be necessary for reproducing the localized adherence phenotype in a laboratory *E. coli* strain. Other investiga- tors using a different approach reported the identical sequence from an unrelated EPEC strain (SOHEL et al. 1993). The deduced amino acid sequence of bundlin confirms that BFP belong to the type-IV family of pili. Members of this family include pili from *Pseudomonas aeruginosa*, *Neisseria gonorrhoeae*, *Vibrio cholerae*, and several other pathogenic and non pathogenic gram-negative spe- cies (HOBBS and MATTICK 1993). Like other members of the family, bundlin is processed from a precursor by a specific prepilin peptidase that removes a characteristic prepilin leader sequence. This enzyme is encoded by the *bfpP* gene, located 7.5 kb downstream of *bfpA* (ZHANG et al. 1994). The EPEC prepilin peptidase shows approximately 30% sequence identity with similar enzymes from other species that produce type-IV fimbriae. The BfpP protein shows reciprocal functional homology with the enzyme from *P. aeruginosa*. The EPEC enzyme can functionally complement a *P. aeruginosa* strain with a mutation in its prepilin peptidase gene, and *P. aeruginosa* can process pre-bundlin.

Attempts to complement *bfpA*::Tn*phoA* mutants revealed that the minimum fragment required to restore localized adherence is 11.2 kb (K.D. Stone and M.S. Donnenberg unpublished data). In contrast, to confer localized adherence upon a laboratory strain that carries the second region of Nataro et al. (1987) required for adherence (see above), a fragment containing an additional 1.6 kb downstream of *bfpA* is needed. In total, as many as 24 kb may be required for the localized adherence phenotype. The DNA sequence of the 12.8-kb fragment reveals the presence of 14 contiguous open reading frames (ORFs) including *bfpA* and *bfpP*. While several of these ORFs resemble genes previously associated with biogenesis of type-IV fimbria, at least three are novel. It is not yet known whether all of these genes encode proteins necessary for synthesis and function of BFP and what, if any, role each product plays. Furthermore, the only gene known to be encoded by the other fragment necessary for localized adherence is a positive regulator of EPEC virulence genes, including *bfpA* (see Sect. 2.5). It is not yet clear whether this regulator alone is sufficient to confer localized adherence in conjunction with the 14 *bfp* genes. It seems likely that several as yet undefined genes encoded on the EPEC adherence plasmid adjacent to the *bfp* cluster are required for localized adherence. Further analysis of these genes and their products should improve our understanding of type-IV fimbrial biogenesis.

2.1.2 The *dsbA* Locus

Analysis of the EPEC sequences adjacent to the sites of transposon insertion in the other category of noninvasive mutants deficient in localized adherence revealed that these mutants have disruptions of the *dsbA* locus (H.-Z. Zhang and M.S. Donnenberg unpublished data). This gene encodes a periplasmic enzyme required for efficient disulfide bond formation in a variety of proteins (Bardwell et al. 1991; Kamitani et al. 1992). Phenotypes associated with mutations of *dsbA* or homologues in other species include reduced activity of alkaline phosphatase (Bardwell et al. 1991; Kamitani et al. 1992), defective assembly of flagella (Dailey and Berg 1993), reduced transcription of outer membrane proteins (Pugsley 1993), reduced adherence and cholera toxin secretion in *V. cholerae* (Peek and Taylor 1992; Yu et al. 1992), absence of transformation in *Hemophilus influenzae* (Tomb 1992), and reduced secretion of cellulase in *Erwinia spp.* (Bortoli-German et al. 1994). In addition, the ability to make F pili (Bardwell et al. 1991) and P fimbriae, but not type-I fimbriae (Jacob-Dubuisson et al. 1994), is lost in *dsbA* mutants. In *V. cholerae*, the adherence phenotype associated with its type-IV fimbria is lost when the *dsbA* homologue is mutated. However, the pili are still produced (Peek and Taylor 1992).

The loss of localized adherence in EPEC *dsbA* mutants suggests that some protein involved in BFP biosynthesis or function has a periplasmic phase of export (H.-Z. Zhang and M.S. Donnenberg unpublished data). This is a matter of importance, in that no periplasmic stage of export has been described for type-IV fimbriae. One candidate for a DsbA substrate is bundlin itself. Like most pilus subunits, bundlin possesses two cysteine residues toward its carboxyl terminus that are predicted to form a disulfide loop. To test whether this bond is necessary for BFP function, cysteine for serine missense mutations were engineered into the *bfpA* gene and exchanged for the wild-type allele in the EPEC background. Both Cys→Ser mutants were found to be deficient in localized adherence, confirming the importance of the disulfide loop in function and supporting the hypothesis that bundlin could be a DsbA substrate (H.-Z.ZHANG and M.S. DONNENBERG unpublished data). Further experiments will determine whether bundlin can act as a DsbA substrate in vitro.

2.2 Intimate Attachment

2.2.1 Attaching and Effacing

The hallmark of EPEC infection is the ability of the organism to attach intimately to epithelial cells and efface microvilli (Fig. 2c). This interaction was first de-scribed by STALEY et al. (1969), while the term "attaching and effacing" was coined years later by MOON et al. (1983). The bacteria adhere to the epithelial cells at a distance of only 10 nm. Microvilli directly beneath the organisms vesiculate and are lost, while those adjacent to bacteria become elongated (KNUTTON et al. 1987). The epithelial cell embraces the bacterium in cuplike pedestals. Cytoskeletal proteins including actin, talin, ezrin, and α-actinin accumulate in the cytoplasm immediately underlying the adherent bacteria (KNUTTON et al. 1989; FINLAY et al. 1992). By fluorescent microscopy, these cytoskeletal proteins are so concen-trated and localized beneath the bacteria that it almost appears as though the organisms themselves are the source of the signal. However, by confocal microscopy it is clear that the fluorescence emanates from the host cell directly beneath the bacteria (FINLAY et al. 1992). In tissue culture, the pedestals become progressively elongated, resembling the protrusions seen in cells infected with *Listeria monocytogenes* or *Shigella flexneri* except that in the case of EPEC, the bacteria are outside the cell membrane (I. ROSENSHINE and B. FINLAY unpublished data).

The attaching and effacing interaction can be separated into intimate attach-ment and signal transduction events, based on the analysis of mutants deficient in each aspect of the process. The signal transduction events will be discussed in Sect 2.3.

2.2.2 Intimin

The *eaeA* gene was first described by JERSE et al. (1990), who analyzed plasmid-cured Tn*phoA* mutants deficient in attaching and effacing. In addition to these mutants are five *eaeA*::Tn*phoA* mutants among the seven noninvasive mutants

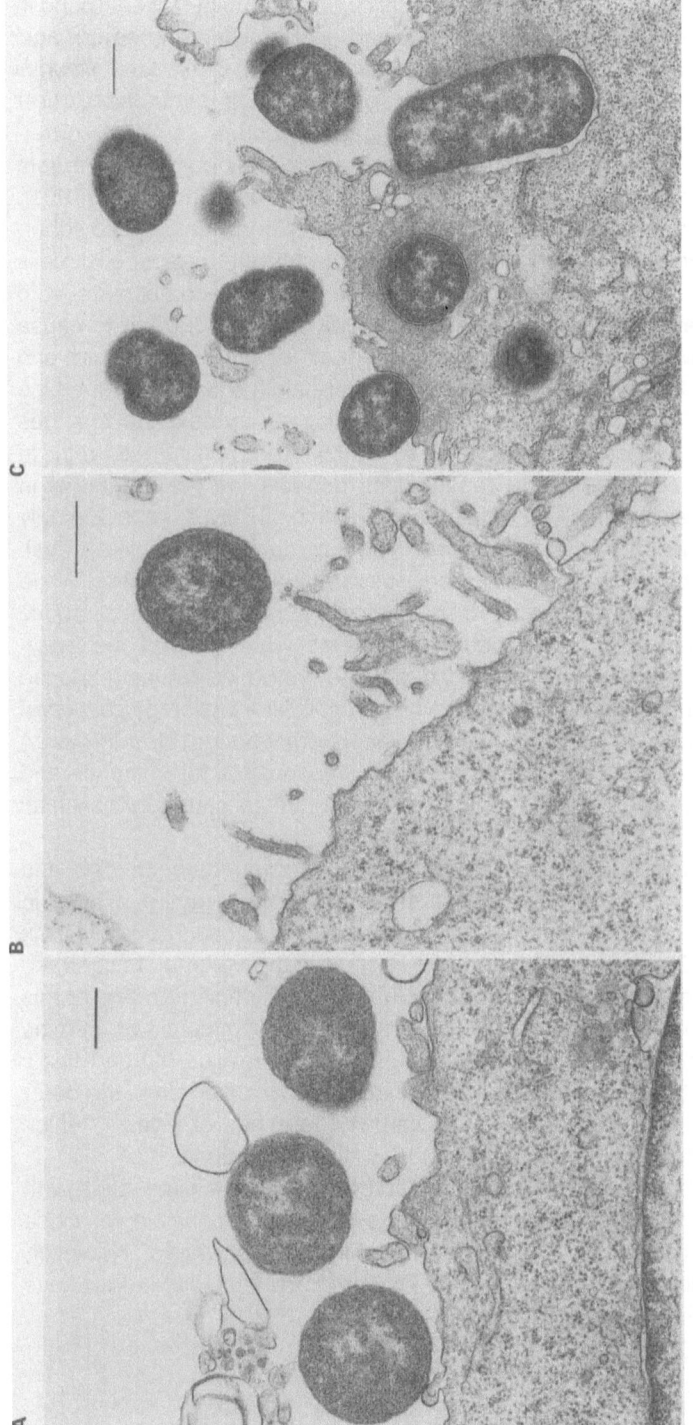

Fig. 2A–C. Electron micrographs demonstrating complementation of attaching and effacing and invasion by co-infection with *eaeA* and *eaeB* mutants. HEp-2 cells were infected with an *eaeA* deletion mutant (**A**), or an *eaeB* deletion mutant (**B**), or co-infected with both mutants (**C**). Only when both mutants are present together are attaching and effacing lesions and intracellular bacteria observed. (Reprinted with permission from FOUBISTER et al. 1994a)

from category 3 (DONNENBERG et al. 1990a). Closely related *eae* genes are found in other bacteria capable of attaching and effacing, including enterohemorrhagic *E. coli* (EHEC) (YU and KAPER 1992; BEEBAKHEE et al. 1992), *Citrobacter freundii* biotype 4280 (SCHAUER and FALKOW 1993a), and some strains of *Hafnia alvei* (ALBERT et al. 1992; FRANKEL et al. 1994). The *eaeA* gene encodes as 94-kD outer-membrane protein known as intimin, that is recongnized by convalescent sera from volunteers experimentally infected with EPEC (JERSE and KAPER 1991). Mutants with transposon insertions or deletions in *eaeA* perform localized adherence normally but are incapable of intimate attachment to epithelial cells (DONNENBERG et al. 1990a; DONNENBERG and KAPER 1991). Recombinants with mutations in similar genes of EHEC and *C.freundii* biotype 4280 fail to cause attaching and effacing lesions in animals (DONNENBERG et a. 1993b; SCHAUER and FALKOW 1993b). Nevertheless, EPEC *eaeA* mutants retain the ability to transduce signals to cells that result in tyrosine kinase activation and cytoskeletal changes (DONNENBERG et al. 1990a; ROSENSHINE et al. 1992). The role of intimin in pathogenesis was confirmed by a randomized, double-blind volunteer trial in which diarrhea developed in all 11 recipients of wild-type EPEC strain, but in only four of 11 recipients of an isogenic *eaeA* deletion mutant (DONNENBERG et al. 1993a). Recipients of the mutant also had less severe illness and less fever. While the mucosal response to infection in the two groups was equivalent, the systemic response was significantly diminished in volunteers who received the *eaeA* mutant. The reduced incidence of fever and reduced systemic immune response could reflect in vivo consequences of diminished cellular invasion by the *eaeA* mutant. The residual diarrhea seen in some volunteers who ingested the *eaeA* mutant strain indicates that intimin is not solely responsible for virulence and correlates well with the retained ability of the mutant to cause cytoskeletal damage in vitro.

Intimin is related to the invasin proteins of *Y. pseudotuberculosis* and *Y. enterocolitica*. Invasin binds avidly to members of the β_1 family of integrin molecules to allow efficient cellular invasion (ISBERG and LEONG 1990; TRAN VAN NHIEU and ISBERG 1993; see also chapter by R.R.ISBERG in this volume). In contrast, intimin, although necessary for invasion by EPEC, is not sufficient to confer this phenotype on noninvasive laboratory *E. coli* strains. The sequences of intimins and invasins are most closely related near the amino terminus (YU and KAPER 1992), a portion of the invasin molecule that is required for insertion in the outer membrane (ISBERG 1989). In contrast, the carboxyl terminus of invasin, which contains the receptor-binding domain, is far less similar to intimin.

Given the phenotype of *eaeA* mutants and the sequence similarities with invasin, intimin is predicted to be the intimate adhesin responsible for close attachment to epithelial cells. Direct evidence for this is limited. However, recombinant proteins consisting of maltose-binding protein fused to intimin have been reported to bind to epithelial cells (FRANKEL et al. 1994).

2.3 Signal Transduction

Cells infected with enteropathogenic *E. coli* undergo profound morphological changes characterized by loss of microvilli, accumulations of cytoskeletal proteins beneath attached bacteria, and rearrangements of the plasma membrane. Ulti-mately, by a process that is poorly understood, net fluid secretion develops culminating in diarrhea. It is evident that the bacteria interrupt or usurp vital signaling pathways in the cell to accomplish these dramatic effects. Although the details of these signaling pathways and the means by which EPEC exploit them remain to be described, progress in identifying some of the steps in signaling has been made. EPEC induce elevations in cytoplasmic calcium and inositol 1, 4, 5-triphosphate levels (BALDWIN et al. 1991; FOUBISTER et al. 1994b; DYTOC et al. 1994). Attaching and effacing lesions can be blocked by inhibitors that prevent the rise in intracellular calcium (DYTOC et al. 1994; BALDWIN et al. 1993). Although *eaeA* mutants, deficient in intimate attachment, fail to cause a rise in intracellular calcium (DYTOC et al. 1994), they do stimulate an inositol phosphate flux (FOUBISTER et al. 1994b). The role of signaling through calcium and inositol phosphate pathways in internalization of EPEC has not been experimentally addressed, but the inositol phosphate flux still occurs when invasion is prevented with cyto-chalasin D, inticating that extracellular EPEC can signal the cell (FOUBISTER et al. 1994b).

2.3.1 Role of Tyrosine Kinases in Internalization

Internalization of EPEC into several cell lines can be inhibited by genestein, a specific tyrosine kinase inhibitor (ROSENSHINE et al. 1992). This result suggests that EPEC invasion is dependent upon signaling through tyrosine kinases. Fur-thermore, the inositol phosphate flux induced by EPEC can be blocked by genestein, suggesting that EPEC-induced tyrosine kinase activation is proximal to the inositol phosphate rise in the signaling cascade (FOUBISTER et al. 1994b). The direct participation of tyrosine kinase signaling in induction of the attaching and effacing effect is suggested by co-localization of tyrosine phosphorylated proteins with cytoskeletal proteins directly beneath attached bacteria (ROSENSHINE et al. 1992). The identity of the specific tyrosine kinase or kinases activated by EPEC is unknown, and the tyrosine kinase substrates phosphorylated in response to EPEC have not been well characterized, but monoclonal antibodies against phosphotyrosine recognize proteins of 90 kD, 72 kD, and 39 KD in epithelial cells infected with EPEC (ROSENSHINE et al. 1992). Similar proteins are recognized in cells infected with an EPEC strain cured of its adherence plasmid and in cells infected with *eaeA* mutants of EPEC, observations which indiĉate that neither BFP nor intimin are required for tyrosine kinase activation. However, the two mutants from the fourth category of non invasive Tn*phoA* mutants fail to induce tyrosine phosphorylation of host cell proteins. These same mutants fail to induce accumulation of filamentous actin beneath the sites of bacterial attachment (DONNENBERG et al. 1990a) and fail to induce an inositol phosphate flux (FOUBISTER et al. 1994b). In contrast, *eaeA* mutants retain the ability to induce accumulations of actin. Thus failure to stimulate tyrosine kinase activation is correlated with inability to cause cytoskeletal disruptions.

Category-4 mutants and *eaeA* mutants are defective in different stages of invasion. Intimin is involved in the intimate attachment of the bacterium to the epithelial cell, while the products of the genes disrupted in category-4 mutants are involved in tyrosine kinase-mediated signaling. To examine the possibility that category-4 mutants and *eaeA* mutants could trans-complement each other, epithelial cells were simultaneously infected with both types of mutants and the invasive ability of each mutant was measured. When co-infected with an *eaeA* mutant, either category-4 mutant is able to invade HeLa cells at nearly wild-type levels (Rosenshine et al. 1992). However, the *eaeA* mutants remain severely deficient for invasion in the presence of either category-4 mutant. This result can be explained as follows: The category-4 mutant, which is able to produce intimin, can come into close contact with the epithelial cell but is unable to signal the cell. When an *eaeA* mutant is present, it signals the cell to allow internalization of the category-4 mutant. In contrast, the *eaeA* mutant, positioned at a distance from the cell, is unable to enter, even in the presence of the category-4 mutant. This unidirectional transcomplementation suggests that a soluble factor, either produced by the *eaeA* mutant or induced in the epithelial cell by this mutant, is able to diffuse to the site of the attached category-4 mutant to allow entry.

2.3.2 A Putative Secretory Apparatus

The genetic defect responsible for the lack of tyrosine kinase-mediated cell signaling by one of the category-4 mutants has recently been described (Jarvis et al. 1995). Although this mutant has two Tn*phoA* insertions, one insertion is in a region of the EPEC chromosome common to K-12 *E. coli*, while the other insertion is in a region absent in K-12 strains. The DNA sequence near the site of the insertion specific to EPEC revealed several genes with predicted products similar to products of virulence loci from other plant and animal pathogens. These products include the Spa and Inv proteins of *Salmonella typhimurium* (Galán et al. 1992, Ginocchio et al. 1992, Groisman and Ochman 1993), the Spa and Mxi proteins of *S. flexneri* (Andrews et al. 1991; Andrews and Maurelli 1992; Venkatesan et al. 1992), and some of the Lcr and Ysc proteins of *Y. enterocolitica* (Plano et al. 1991; Michiels et al. 1991). All of these proteins are necessary for the secretion of soluble virulence factors, such as the Ipa proteins of *Shigella* (Mènard et al. 1993) and the Yops of *Yersinia* (Straley et al. 1993), and are hypothesized to form a secretion apparatus that exports such proteins beyond the outer membrane of the bacteria. Thus far, four such genes have been described in EPEC and denoted *sepA–D* (Jarvis et al. 1995). The site of the Tn*phoA* insertion in one of the category-4 mutants has been localized to *sepA*. An additional mutant, engineered to contain a kanamycin-resistance gene insertion in the *sepB* gene, is deficient in inducing cytoskeletal alterations in tissue culture cells and fails to induce phosphorylation of the 90-kD host cell tyrosine kinase substrate.

If EPEC possess a secretion apparatus, what proteins are secreted? KENNY and FINLAY and JARVIS et al. independently reported several proteins in supernatants of EPEC grown in tissue culture medium (JARVIS et al. 1995; KENNY and FINLAY 1995). Both groups found that category-4 mutants fail to export most of the secreted proteins identified. In addition, the *sepB* mutant fails to export these proteins (JARVIS et al. 1995). Thus EPEC mutants deficient in signal transduction have mutations in genes proposed to be involved in secretion of several proteins. These secreted proteins are candidates for effector molecules that initiate tyrosine kinase-dependent signal transduction events in the host cell.

2.3.3 EaeB, a Secreted Protein Necessary for Intimate Attachment and Signal Transduction

Located 5 kb downstream of the *eaeA* gene is a locus known as *eaeB* that is required for attaching and effacing activity. The *eaeB* gene was identified by analyzing plasmids required to restore attaching and effacing ability to the two non invasive mutants from category three that have Tn*phoA* insertions downstream of *eaeA*. A mutant engineered to have an in-frame deletion within *eaeB* is not only incapable of intimate attachment to epithelial cells like the *eaeA* mutant (DONNENBERG et al. 1993c), but is also deficient in tyrosine kinase-mediated signal transduction and fails to cause an inositol phosphate flux, like category-4 mutants (FOUBISTER et al. 1994a). Both intimate attachment and signal transduction activities are restored by reintroduction of the *eaeB* gene to the mutant on a plasmid. Furthermore, co-infection of epithelial cells with *eaeA* and *eaeB* deletion mutants restores wild-type attaching and effacing lesions and enhances invasion (Fig. 2). This transcomplementation, like that between *eaeA* mutants and category-4 mutants, is unidirectional. Only internalization of the *eaeB* mutant is increased by co-infection; the *eaeA* mutant remains outside the cell (FOUBISTER et al. 1994a).

Amino acid sequencing of proteins that are secreted by wild-type EPEC but not category-4 mutants revealed that one of these proteins is EaeB (KENNY and FINLAY 1995). The discovery that the EaeB product is one of the *sep*-dependent secreted proteins helps to clarify the transcomplementation results. Thus EaeB is secreted, presumably by the apparatus encoded by the *sep* genes, and is capable of diffusion from the *eaeA* mutant either to the *eaeB* mutant or to the epithelial cell. Once supplied with the function of the EaeB protein, the *eaeB* mutant becomes capable of signal transduction, intimate attachment, and invasion.

The precise function and target of the EaeB protein are unknown. Recently a gene from *S.typhimurium* that encodes a secreted protein with limited sequence similarities to EaeB has been identified (COLLAZO et al. 1995). Mutations in the *Salmonella* locus eliminate tyrosine kinase-mediated signal transduction by that species as well. The sequence of the *eaeB* gene predicts a protein that contains a pyridoxal phosphate-binding site, but binding of this cofactor has not been demonstrated experimentally.

2.4 The Locus of Enterocyte Effacement

The *eaeA* gene, the *eaeB* gene, and the *sep* genes all map to a single *Notl* restriction fragment by pulse-field gel electrophoresis (McDANIEL et al. 1995). Further mapping using cosmid clones has revealed that all of these loci reside on a common 35-kb pathogenicity island (Fig. 3). Probes from the entire length of this virulence cassette hybridize to DNA from all EPEC and EHEC tested, as well as other attaching and effacing species, but not to other *E.coli*. The cassette has been termed the locus of enterocyte effacement, or LEE. The sequences adjacent to the LEE hybridize with bacteriophage clones from 82 min of the K-12 chromosome, and the sequence of the boundaries of the LEE reveals that it is inserted at the *selC* gene, which encodes the selenocysteine tRNA. The precise site of insertion is identical to the site of insertion of a pathogenicity island in uropathogenic *E. coli* that includes the genes for hemolysin production. In addition, at least one end of the EPEC and one end of the uropathogenic *E. coli* inserts display remarkable sequence identity (McDANIEL et al. 1995).

In addition to the loci already described, the LEE includes three ORFs between *eaeA* and *eaeB*. The nucleotide sequence of one of these ORFs predicts a protein with an amino terminus identical to that of one of the *sep*-dependent secreted proteins (KENNY and FINLAY 1995). In addition, downstream of *eaeB* are three more kilobases of sequence prior to the junction of the *selC* gene. An omega interposon insertion immediately downstream of *eaeB* disrupts attaching and effacing, suggesting that these sequences include additional genes required for EPEC pathogenesis.

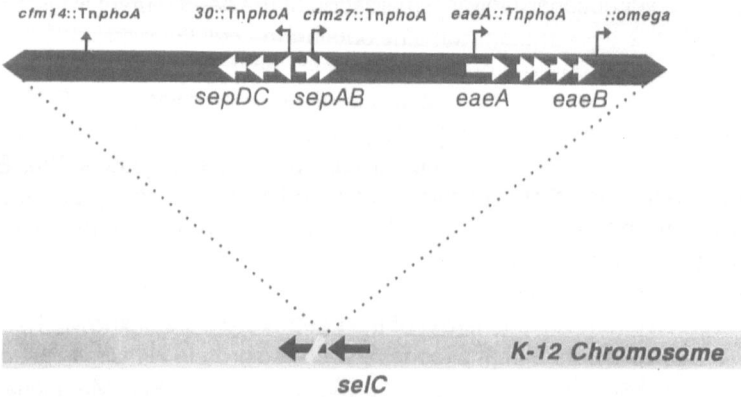

Fig. 3. Map of the locus of enterocyte effacement (LEE). Mutations affecting invasion or attaching and effacing are depicted above the locus, including sites of insertion of transposon Tn*phoA* in category-4 mutants (*cfm14*::Tn*phoA* and *cfm27*::Tn*phoA*) and a category-5 mutant (*30*::Tn*phoA*). Also shown is the site of an omega interposon insertion. *Arrows* within the LEE depict genes or open reading frames, while designations below the fragment show the names of the genes described thus far. The site of insertion of the LEE relative to the K-12 genome at the *selC* gene is indicated by the *broken lines*. Downstream of *selC* is an open reading frame of unknown function that has undergone a deletion, indicated by the *hash mark,* in EPEC strains

Upstream of *eaeA* in the LEE are the *sep* loci. Between *sepB* and *sepC* is a small gene that has not yet been named. It is the site of Tn*phoA* insertion in a noninvasive mutant from the fifth category. This mutant is capable of localized adherence and intimate attachment (R.P.RABINOWITZ and M.S. DONNENBERG unpublished data). It forms attaching and effacing lesions that, by electron microscopy, are quantitatively and qualitatively indistinguishable from the wild type. Yet this mutant is severely defective in its ability to invade epithelial cells. Further characterization of this mutant is in progress.

Finally, near the end of the LEE, opposite to *eaeB*, is the site of one of the two Tn*phoA* inserts from the other category-4 mutant (MCDANIEL et al. 1995). The gene into which the transposon has inserted in this mutant has not yet been characterized.

2.5 Regulation of EPEC Virulence Factors

The expression of several EPEC virulence factors involved in invasion is responsive to culture conditions. Growth in tissue culture medium enhances expression of intimin (JERSE and KAPER 1991), BFP (DONNENBERG et al. 1992; VUOPIO-VARKILA and SCHOOLNIK 1991), and the secreted proteins including EaeB (JARVIS et al. 1995; KENNY and FINLAY 1995). The increased expression of all of these proteins under the same culture conditions suggests that EPEC virulence factors are coordinately regulated.

The ability of plasmid-cured strains of EPEC to invade tissue culture cells is reduced in comparison to that of strains carrying the EPEC adherence plasmid (DONNENBERG et al. 1989; FRANCIS et al. 1991). The demonstration that Tn*phoA* insertions in the *bfpA* gene eliminate localized adherence (DONNENBERG et al. 1992) and reduce invasion (DONNENBERG et al. 1990a) provides a partial explanation for this observation. There may be an additional explanation as well. The presence of the adherence plasmid increases expression of the *eaeA* gene (JERSE and KAPER 1991). A fragment containing a putative positive regulatory factor has been subcloned from the adherence plasmid (GÓMEZ 1994). In the presence of a plasmid containing the cloned factor, expression of *eaeA*, *eaeB*, and *bfpA* are all increased at the transcriptional level. The locus responsible has been called *per* for Plasmid-encoded regulator. Further characterization of *per* is in progress.

3 Other Internalization Systems

FLETCHER et al. (1992) reported that a fragment subcloned from a plasmid found in an 0111:H- EPEC strain is capable of directing invasion of a laboratory *E. coli* strain. The plasmid from which the fragment was subcloned is distinct from the EPEC adherence plasmid that contains *bfbA*, which is also present in the EPEC strain used. The 4.5-kb fragment, when cloned in a moderate copy number vector, directs invasion of HEp-2 cells at a level that is 1000-fold higher than the laboratory strain carrying the vector alone, but this level is also 30-fold higher than the laboratory strain carrying the entire plasmid from which the fragment was subcloned. This difference suggests that the copy number of the subcloned locus may influence its ability to direct invasion. DNA hybridization using the colned fragment revealed that it is present in a minority of EPEC strains. Further characterization of the fragment has not been reported.

4 Summary and Conclusions

EPEC are capable of efficient invasion into a variety of cell lines in vitro. The significance of this cellular invasion in the development of disease is not known, but the analysis of mutants deficient in the process has allowed the dissection of EPEC cellular pathogenesis into three distinct stages. These stages are coordinated by a plasmid-encoded regulator and may occur simultaneously. Initial adherence is associated with the production of a type-IV fimbria called the bundle-forming pilus. More than 14 genes encoded on the large EPEC palsmid, as well as a chromosomal gene encoding a periplasmic disulfide bond-forming enzyme, are required for biogenesis and function of the bundle-forming pilus. The second stage consists of signal transduction events directed by the bacterium, which include activation of host cell tyrosine kinase activity, resulting in inositol phosphate and calcium fluxes and disruptions of the host cell cytoskeleton. EPEC secrete several proteins, including the product of the *eaeB* gene which is required for the signal transduction events. Secretion of these proteins is dependent upon genes that resemble loci with a similar function in other invasive enteric pathogens. The third stage, intimate attachment of EPEC to the host cell, requires intimin, an outer-membrane protein related to the *Yersinia* invasins. The genes encoding the putative secretory apparatus, intimin, EaeB, and other products required for invasion are located within a contiguous 35-kb locus of enterocyte effacement that is inserted at a specific site relative to the K-12 *E. coli* genome in many EPEC strains.

Further studies of the genes and gene products required for EPEC invasion are likely to clarify EPEC pathogenesis. Priorities for future work include identifying the effector molecules of EPEC signal transduction and the targets of these factors, identifying receptors for the putative EPEC ligands intimin and the bundle-forming pilus, and elucidating the cascade that ensues in the target cell and results in fluid secretion. Further analysis of interactions between EPEC and host cells is likely to have broad implications for the understanding of bacterial pathogenesis.

Acknowledgments. The author is grateful to Jim Kaper, Karen Jarvis, Tim McDaniel, Brendan Kenny, Brett Finlay, Ilan Rosenshine, and Jorge Galán for communicating results prior to publication. Work conducted in the author's laboratory was supported by Public Service Award AI32074 from the National Institutes of Health and by the Department of Veterans's Affairs.

References

Albert MJ, Faruque SM, Ansaruzzaman M, Islam MM, Haider K, Alam K, Kabir I, Robins-Browne R (1992) Sharing of virulence-associated properties at the phenotypic and genetic levels between enteropathogenic *Escherichia coli and Hafnia alvei*.. J Med Microbiol 37: 310–314

Andrade JR, Da Veiga VF, De Santa Rosa MR, Suassuna I (1989) An endocytic process in HEp-2 cells induced by enteropathogenic *Escherichia coli*. J Med Microbiol 28: 49–57

Andrews GP, Maurelli AT (1992) *mxiA* of *Shigella flexneri* 2a, which facilitates export of invasion plasmid antigens, encodes a homolog of the low-calcium-response protein, LcrD, of *Yersinia pestis*. Infect Immun 60: 3287–3295

Andrews GP, Hromockyj AE, Coker C, Maurelli AT (1991) Two novel virulence loci, *mxiA and mxiB*, in *Shigella flexneri* 2a facilitate excretion of invasion plasmid antigens. Infect Immun 59: 1997–2005

Baldini MM, Kaper JB, Levine MM, Candy DC, Moon HW (1983) Plasmid-mediated adhesion in enteropathogenic *Escherichia coli*. J Pediatr Gastroenterol Nutr 2: 534–538

Baldwin TJ, Ward W, Aitken A, Knutton S, Williams PH (1991) Elevation of intracellular free calcium levels in HEp-2 cells infected with enteropathogenic *Escherichia coli*. Infect Immun 59: 1599–1604

Baldwin TJ, Lee-Delaunay MB, Knutton S, Williams PH (1993) Calcium-calmodulin dependence of actin accretion and lethality in cultured HEp-2 cells infected with enteropathogenic *Escherichia coli*. Infect Immun 61: 760–763

Bardwell JC, McGovern K, Beckwith J (1991) Identification of a protein required for disulfide bond formation in vivo. Cell 67: 581–589

Beebakhee G. Louie M, De Azavedo J, Brunton J (1992) Cloning and nucleotide sequence of the *eae* gene homologue from enterohemorrhagic *Escherichia coli* serotype 0157: H7. FEMS Microbiol Lett 91: 63–68

Bortoli-German I, Brun E, Py B, Chippaux M, Barras F (1994) Periplasmic disulphide bond formaiton is essential for cellulase secretion by the plant pathogen *Erwinia chrysanthemi*. Mol Microbiol 11: 545–553

Bower JR, Congeni BL, Cleary TG, Stone RT, Wanger A, Murray BE, Mathewson JJ, Pickering LK (1989) *Escherichia coli* 0114: nonmotile as a pathogen in an outbreak of severe diarrhea associated with a day care center. J Infect Dis 160: 243–247

Canil C, Rosenshine I, Ruschkowski S, Donnenberg MS, Kaper JB, Finlay BB (1993) Enteropathogenic *Escherichia coli* decreases the transepithelial electrical resistance of polarized epithelial monolayers. Infect Immun 61: 2755–2762

Cobeljić M, Mel D, Arsić B, Krstić L, Sokolovski B, Nikolovski B, Sopovski E, Kulauzov M, Kalenić S (1989) The association of enterotoxigenic and enteropathogenic *Escherichia coli* and other enteric pathogens with childhood diarrhoea in Yugoslavia. Epidemiol Infect 103: 53–62

Collazo CM, Zierler MK, Galán JE (1995) Functional analysis of the *Salmonella typhimurium* invasion genes *invI* and *invJ* and identification of a target of the protein secretion apparatus encoded in the *inv* locus. Mol Microbiol 15: 25–38

Cravioto A, Gross RJ, Scotland SM, Rowe B (1979) An adhesive factor found in strains of *Escherichia coli* belonging to the traditional infantile enteropathogenic serotypes. Current Microbiol 3: 95–99

Cravioto A, Reyes R, Ortega R, Fernández G, Hernández R, López D (1988) Prospective study of diarrhoeal disease in a cohort of rural Mexican children: incidence and isolated pathogens during the first two years of life. Epidemiol Inf etc 101: 123–134

Cravioto A, Tello A, Navarro A, Ruiz J, Villafán H, Uribe F, Eslava C (1991) Association of *Escherichia coli* HEp-2 adherence patterns with type and duration of diarrhoea. Lancet 337: 262–264

Dailey FE, Berg HC (1993) Mutants in disulfide bond formation that disrupt flagellar assembly in *Escherichia coli*. Proc Natl Acad Sci USA 90: 1043–1047

Donnenberg MS, Kaper JB (1991) Construction of an *eae* deletion mutant of enteropathogenic *Escherichia coli* by using a positive-selection suicide vector. Infect Immun 59: 4310–4317

Donnenberg MS, Kaper JB (1992) Minireview: enteropathogenic *Escherichia coli*. Infect Immun 60: 3953–3961

Donnenberg MS, Donohue-Rolfe A, Keusch GT (1989) Epithelial cell invasion: an overlooked property of enteropathogenic *Escherichia coli* (EPEC) associated with the EPEC adherence factor. J Infect Dis 160: 452–459

Donnenberg MS, Calderwood SB, Donohue-Rolfe A, Keusch GT, Kaper JB (1990a) Construction and analysis of Tn*phoA* mutants of enteropathogenic *Escherichia coli* unable to invade HEp-2 cells. Infect Immun 58: 1565–1571

Donnenberg MS, Donohue-Rolfe A, Keusch GT (1990b) A comparison of HEp-2 cell invasion by enteropathogenic and enteroinvasive *Escherichia coli*. FEMS Microbiol Lett 57: 83–86

Donnenberg MS, Girón JA, Nataro JP, Kaper JB (1992) A plasmid-encoded type IV fimbrial gene of enteropathogenic *Escherichia coli* associated with localized adherence. Mol Microbiol 6: 3427–3437

Donnenberg MS, Tacket Co, James SP, Losonsky G, Nataro JP, Wasserman SS, Kaper JB, Levine MM (1993a) The role of the *eaeA* gene in experimental enteropathogenic *Escherichia coli* infection. J Clin Invest 92: 1412–1417

Donnenberg MS, Tzipori S, McKee M, O'Brien AD, Alroy J, Kaper JB (1993b) The role of the *eae* locus of enterohemorrhagic *Escherichia coli* in intimate attachment in vitro and in a porcine model. J Clin Invest 92: 1418–1424

Donnenberg MS, Yu J, Kaper JB (1993c) A second chromosomal gene necessary for intimate attachment of enteropathogenic *Escherichia coli* to epithelial cells. J Bacteriol 175: 4670–4680

Dytoc M, Fedorko L, Sherman PM (1994) Signal transduction in human epithelial cells infected with attaching and effacing *Escherichia coli* in vitro. Gastroenterol 106: 1150–1161

Echeverria P, Orskov F, Orskov I, Knutton S, Scheutz F, Brown JE, Lexomboon U (1991) Attaching and effacing enteropathogenic *Escherichia coli* as a cause of infantile diarrhea in Bangkok. J Infect Dis 164: 550–554

Fagundes Neto U, Ferreira V, Patricio FRS, Mostaço VL, Trabulsi LR (1989) Protracted diarrhea: the importance of the enteropathogenic *E. coli* (EPEC) strains and *Salmonella* in its genesis. J Pediatr Gastroenterol Nutr 8: 207–211

Finlay BB, Rosenshine I, Donnenberg MS, Kaper JB (1992) Cytoskeletal composition of attaching and effacing lesions associated with enteropathogenic *Escherichia coli* adherence to HeLa cells. Infect Immun 60: 2541–2543

Fletcher JN, Embaye HE, Getty B, Batt RM, Hart CA, Saunders JR (1992) Novel invasion determinant of enteropathogenic *Escherichia coli* plasmid pLV501 encodes the ability to invade intestinal epithelial cells and HEp-2 cells. Infect Immun 60: 2229–2236

Foubister V, Rosenshine I, Donnenberg MS, Finlay BB (1994a) The *eaeB* gene of enteropathogenic *Escherichia coli* is necessary for signal transduction in epithelial cells. Infect Immun 62: 3038–3040

Foubister V, Rosenshine I, Finlay BB (1994b) A diarrheal pathogen, enteropathogenic *Escherichia coli* (EPEC), triggers a flux of inositol phosphates in infected epithelial cells. J Exp Med 179: 993–998

Francis CL, Jerse AE, Kaper JB, Falkow S (1991) Characterization of interactions of enteropathogenic *Escherichia coli* 0127: H6 with mammalian cells in vitro. J Infect Dis 164: 693–703

Frankel G, Candy DCA, Everest P, Dougan G (1994) Characterization of the C-terminal domains of intimin-like proteins of enteropathogenic and enterohemorrrhagic *Escherichia coli*, *Citrobacter freundii*, and *Hafnia alvei*. Infect Immun 62: 1835–1842

Galán JE, Ginocchio C, Costeas P (1992) Molecular and functional characterization of the *Salmonella* invasion gene *invA*: homology of InvA to members of a new protein family. J Bacteriol 174: 4338–4349

Giles C, Sangster G, Smith J (1949) Epidemic gastroenteritis of infants in Aberdeen during 1947. Arch Dis Child 24: 45–53

Ginocchio C, Pace J, Galán JE (1992) Identification and molecular characterization of a *Salmonella typhimurium* gene involved in triggering the internalization of salmonellae into cultured epithelial cells. Proc Natl Acad Sci USA 89: 5976–5980

Girón JA, Ho ASY, Schoolnik GK (1991) An inducible bundle-forming pilus of enteropathogenic *Escherichia coli*. Science 254: 710–713

Gomes TAT, Rassi V, Macdonald KL, Ramos SRTS, Trabulsi LR, Vieira MAM, Guth BEC, Candeias JAN, Ivey C, Toledo MRF, Blake PA (1991) Enteropathogens associated with acute diarrheal disease in urban infants in São Paulo, Brazil. J Infect Dis 164: 331–337

Gómez-Duarte OG, Kaper JB (1995) A plasmid-encoded regulatory region activates chromosomal *eaeA* expression in enteropathogenic *Escherichia coli*. Infect Immun 63: 1767–1776

Groisman EA, Ochman H (1993) Cognate gene clusters govern invasion of host epithelial cells by *Salmonella typhimurium* and *Shigella flexneri*. EMBO J 12: 3779–3787

Hill SM, Phillips AD, Walker-Smith JA (1991) Enteropathogenic *Escherichia coli* and life-threatening chronic diarrhoea. Gut 32: 154–158

Hobbs M, Mattick JS (1993) Common components in the assembly of type 4 fimbriae, DNA transfer systems, filamentous phage and protein-secretion apparatus: a general system for the formation of surface-associated protein complexes. Mol Microbiol 10: 233–243

Isberg RR (1989) Mammalian cell adhesion functions and cellular penetration of enteropathogenic *Yersinia* species. Mol Microbiol 3: 1449–1453

Isberg RR, Leong JM (1990) Multiple β_1 chain integrins are receptors for invasin, a protein that promotes bacterial penetration into mammalian cells. Cell 60: 861–871

Jacob-Dubuisson F, Pinkner J, Xu Z, Striker R, Padmanhaban A, Hultgren SJ (1994) PapD chaperone function in pilus biogenesis depends on oxidant and chaperone-like activities of DsbA. Proc Natl Acad Sci USA 91: 11552–11556

Jarvis KG, Girón JA, Jerse AE, McDaniel TK, Donnenberg MS, Kaper JB (1995) Enteropathogenic *Escherichia coli* contains a specialized secretion system necessary for the export of proteins involved in attaching and effacing lesion formation. Proc Natl Acad Sci USA 92: 7996–8000

Jerse AE, Yu J, Tall BD, Kaper JB (1990) A genetic locus of enteropathogenic *Escherichia coli* necessary for the production of attaching and effacing lesions on tissue culture cells. Proc Natl Acad Sci USA 87: 7839–7843

Jerse AE, Kaper JB (1991) The *eae* gene of enteropathogenic *Escherichia coli* encodes a 94-kilodalton membrane protein, the expression of which is influenced by the EAF plasmid. Infect Immun 59: 4302–4309

Kain KC, Barteluk RL, Kelly MT, Xin H, Hua GD, Yuan H, Proctor EM, Byrne S, Stiver HG (1991) Etiology of childhood diarrhea in Beijing, China. J Clin Microbiol 29: 90–95

Kamitani S, Akiyama Y, Ito K (1992) Identification and characterization of an *Escherichia coli* gene required for the formation of correctly folded alkaline phosphatase, a periplasmic enzyme. EMBO J 11: 57–62

Kenny B, Finlay BB (1995) Secretion of proteins by enteropathogenic *E. coli* which mediate signaling in host epithelial cells. Proc Natl Acad Sci USA 92: 7991–7995

Knutton S, Lloyd DR, McNeish AS (1987) Adhesion of enteropathogenic *Escherichia coli* to human intestinal enterocytes and cultured human intestinal mucosa. Infect Immun 55: 69–77

Knutton S, Baldwin T, Williams PH, McNeish AS (1989) Actin accumulation at sites of bacterial adhesion to issue culture cells: basis of a new diagnostic test for enteropathogenic and entero-hemorrhagic *Escherichia coli*. Infect Immun 57: 1290–1298

Levine MM, Prado V, Robins-Browne R, Lior H, Kaper JB, Moseley SL, Gicquelais K, Nataro JP, Vial P, Tall B (1988) Use of DNA probes and HEp-2 cell adherence assay to detect diarrheagenic *Escherichia coli*. J Infect Dis 158: 224–228

McDaniel TK, Jarvis KG, Donnenberg MS, Kaper JB (1995) A genetic locus of enterocyte effacement conserved among diverse enterobacterial pathogens. Proc Natl Acad Sci USA 92: 1664–1668

Ménard R, Sansonetti PJ, Parsot C (1993) Nonpolar mutagenesis of the *ipa* genes defines IpaB, IpaC, and IpaD as effectors of *Shigella flexneri* entry into epithelial cells. J Bacteriol 175: 5899–5906

Michiels T, Vanooteghem J-C, DeRouvroit CL, China B, Gustin A, Boudry P, Cornelis GR (1991) Analysis of *virC*, an operon involved in the secretion of Yop proteins by *Yersinia enterocolitica*. J Bacteriol 173: 4994–5009

Miliotis MD, Koornhof HJ, Phillips JI (1989) Invasive potential of noncytotoxic enteropathogenic *Escherichia coli* in an in vitro Henle 407 cell model. Infect Immun 57: 1928–1935

Moon HW, Whipp SC, Argenzio RA, Levine MM, Giannella RA, (1983) Attaching and effacing activities of rabbit and human enteropathogenic *Escherichia coli* in pig and rabbit intestines. Infect Immun 41: 1340–1351

Nataro JP, Maher KO, Mackie P, Kaper JB (1987) Characterization of plasmids encoding the adherence factor of enteropathogenic *Escherichia coli*. Infect Immun 55: 2370–2377

Paulozzi LJ, Johnson KE, Kamahele LM, Clausen CR, Riley LW, Helgerson SD (1986) Diarrhea associated with adherent enteropathogenic *Escherichia coli* in an infant and toddler center, Seattle, Washington. Pediatrics 77: 296–300

Peek JA, Taylor RK (1992) Characterization of a periplasmic thiol: disulfide interchange protein required for the functional maturation of secreted virulence factors of *Vibrio cholerae*. Proc Natl Acad Sci USA 89: 6210–6214

Plano GV, Barve SS, Straley SC (1991) LcrD, a membrane-bound regulator of the *Yersinia pestis* low-calcium response. J Bacteriol 173: 7293–7303

Polotsky YE, Dragunskaya EM, Seliverstova VG, Avdeeva TA, Chakhutinskaya MG, Kétyi I, Vertényi A, Ralovich B, Emödy L, Málovics I, Safonova NV, Snigirevskaya ES, Karyagina EI (1977) Pathogenic effect of enterotoxigenic *Escherichia coli* and *Escherichia coli* causing infantile diarrhoea. Acta Microbiol Acad Sci Hung 24: 221–236

Pugsley AP (1993) A mutation in the *dsbA* gene coding for periplasmic disulfide oxidoreductase reduces transcription of the *Escherichia coli ompF* gene. Mol Gen Genet 237: 407–411

Riley LW, Junio LN, Schoolnik GK (1990) HeLa cell invasion by a strain of enteropathogenic *Escherichia coli* that lacks the O-antigenic polysaccharide. Mol Microbiol 4: 1661–1666

Robins-Browne RM, Bennett-Wood V (1992) Quantitative assessment of the ability of *Escherichia coli* to invade cultured animal cells. Microb Pathog 12: 159–164

Robins-Browne R, Still CS, Miliotis MD, Richardson NJ, Koornhof HJ, Frieman I, Schoub BD, Lecatsas G, Hartman E (1980) Summer diarrhoea in African infants and children. Arch Dis Child 55: 923–928

Rosenshine I, Donnenberg MS, Kaper JB, Finlay BB (1992) Signal exchange between entero-pathogenic *Escherichia coli* (EPEC) and epithelial cells: EPEC induce tyrosine phosphorylation of host cell protein to initiate cytoskeletal rearrangement and bacterial uptake. EMBO J 11: 3551–3560

Scaletsky ICA, Silva MLM, Trabulsi LR (1984) Distinctive patterns of adherence of enteropathogenic *Escherichia coli* to HeLa cells. Infect Immun 45: 534–536

Schauer DB, Falkow S (1993a) Attaching and effacing locus of a *Citrobacter freundii* biotype that causes transmissible colonic hyperplasia. Infect Immun 61: 2486–2492

Schauer DB, Falkow S (1993b) The *eae* gene of *Citrobacter freundii* biotype 4280 is necessary for colonization in transmissible murine colonic hyperplasia. Infect Immun 61: 4654–4661

Schtoeder SA, Caldwell JR, Vernon TM, White PS, Granger SI, Bennett JV (1968) A waterborne outbreak of gastroenteritis in adults associated with enteropathogenic *Escherichia coli*. Lancet 1: 737–740

Senerwa D, Olsvik O, Mutanda LN, Lindqvist KJ, Gathuma JM, Fossum K, Wachsmuth K (1989) Enteropathogenic *Escherichia coli* serotype 0111: HNT isolated from preterm neonates in Nairobi, Kenya. J Clin Microbiol 27: 1307–1311

Sohel I, Puente JL, Murray WJ, Vuopio-Varkila J, Schoolnik GK (1993) Cloning and characterization of the bundle-forming pilin gene of enteropathogenic *Escherichia coli* and its distribution in *Salmonella* serotypes. Mol Microbiol 7: 563–575

Staley TE, Jones EW, Corley LD (1969) Attachment and penetration of *Escherichia coli* into intestinal epithelium of the ileum in newborn pigs. Am J Pathol 56: 371–392

Straley SC, Skrzypek E, Plano GV, Bliska JB (1993) Yops of *Yersinia* spp. pathogenic for humans. Infect Immun 61: 3105–3110

Thomb J-F (1992) A periplasmic protein disulfide oxidoreductase is required for transformation of *Haemophilus influenzae* Rd. Proc Natl Acad Sci USA 89: 10252–10256

Tran Van Nhieu G, Isberg RR (1993) Bacterial internalization mediated by β_1 chain integrins is determined by ligand affinity and receptor density. EMBO J 12: 1887–1895

Tzipori S, Gibson R, Montanaro J (1989) Nature and distribution of mucosal lesions associated with enteropathogenic and enterohemorrhagic *Escherichia coli* in piglets and the role of plasmid-mediated factors. Infect Immun 57: 1142–1150

Ulshen MH, Rollo JL (1980) Pathogenesis of *Escherichia coli* gastroenteritis in man—another mechanism. N Engl J Med 302: 99–101

Venkatesan MM, Buysse JM, Oaks EV (1992) Surface presentation of *Shigella flexneri* invasion plasmid antigens requires the products of the *spa* locus. J Bacteriol 174: 1990–2001

Viljanen MK, Peltola T, Junnila SYT, Olkkonen L, Järvinen H, Kuistila M, Huovinen P (1990) Outbreak of diarrhoea due to *Escherichia coli* 0111: B4 in schoolchildren and adults: association of Vi antigen-like reactivity. Lancet 336: 831–834

Vuopio-Varkila J, Schoolnik GK (1991) Localized adherence by enteropathogenic *Escherichia coli* is an inducible phenotype associated with the expression of new outer membrane proteins. J Exp Med 174: 1167–1177

Wu S-X, Peng R-Q (1992) Studies on an outbreak of neonatal diarrhea caused by EPEC 0127: H6 with plasmid analysis restriction analysis and outer membrane protein determination. Acta Paediatr Scand 81: 217–221

Yu J, Webb H, Hirst TR (1992) A homologue of the *Escherichia coli* DsbA protein involved in disulphide bond formation is required for enterotoxin biogenesis in *Vibrio cholerae*. Mol Microbiol 6: 1949–1958

Yu J, Kaper JB (1992) Cloning and characterization of the *eae* gene of enterohemorrhagic *Escherichia coli* 0157: H7. Mol Microbiol 6: 411–417

Zhang H-Z, Lory S, Donnenberg MS (1994) A plasmid-encoded prepilin peptidase gene from entero-pathogenic *Escherichia coli*. J Bacteriol 176: 6885–6891

Legionella pneumophila Invasion of Mononuclear Phagocytes

H.A. Shuman[1], M.A. Horwitz[2]

1 Introduction

Legionella pneumophila is a gram-negative bacterium that causes legionnaires' disease. This organism is widespread in fresh water and is typically found growing in association with protozoans and blue-green algae. In human beings, *L. pneumophila* infects alveolar macrophages, wherein the organism survives and replicates within a specialized phagosome or vacuole (the *Legionella*-specialized phagosome, LSP). The interaction between *L. pneumophila* and human

[1] Department of Microbiology, College of Physicians and Surgeons, Columbia University, 701 West 168th Street, New York, NY 10032, USA
[2] Division of Infectious Diseases, Department of Medicine, School of Medicine, University of California, Los Angeles, 10833 Le Conte Avenue, Los Angeles, CA 90024, USA

mononuclear phagocytes has been studied in considerable detail, and it provides an interesting and informative example of how one organism successfully achieves intracellular parasitism.

L. pneumophila is a facultative intracellular pathogen, but in contrast to many such organisms, L. pneumophila does not grow in tissue culture medium. Consequently, any growth that is observed in cultured cells is due to intracellular multiplication. This facilitates the study of factors that are important exclusively for growth inside host cells and constitutes a significant technical advantage of the Legionella model. In addition, a variety of genetic tools have been developed for Legionella, rendering it readily amenable to genetic analysis.

The central issues which have been under investigation are: How does L.pneumophila gain entry into macrophages? What are the molecular characteristics of the organism's phagosome, and with what host cell organelles does the phagosome interact? How does the organism survive the antimicrobial capacities of the macrophage? Finally, which genes or gene products are critical for intracellular survival and multiplication? The following sections describe results obtained by studying both the cell biology and the genetics of this host-pathogen interaction.

2 Phagocytosis

2.1 Morphology

L. pneumophila enters phagocytes, including monocytes, alveolar macrophages, and polymorphonuclear leukocytes, by a process termed coiling phagocytosis, in which long phagocyte pseudopods coil around the organism as it is internalized (Horwitz 1984). The coiling phenomenon is not unique to L. pneumophila. Other organisms, including the intracellular pathogens Leishmania donovani (Chang 1979) and Chlamydia psittaci (Wyrick and Brownridge 1978) and the pathogens Trypanosoma brucei (Stevens and Moulton 1978) and Borrelia burgdorferi (Szczepanski and Fleit 1988) have also been observed entering phagocytes within a pseudopod coil. The relationship, if any, of the coiling phenomenon to intracellular pathogenesis is not known. While the intracellular pathogens just noted enter by coiling, other intracellular pathogens enter by conventional phagocytosis; these include Mycobacterium tuberculosis (Schlesinger et al. 1990), Mycobacterium leprae (Schlesinger and Horwitz 1990a, b), and Trypanosoma cruzi (Noguira and Cohn 1976; Tanowitz et al. 1975).

Coating L. pneumophila with antibody against the organism neutralizes the coiling phenomenon; such organisms enter phagocytes by conventional phagocytosis (Horwitz 1984). Beyond its effect on the morphology of entry, anti-L. pneumophila antibody in the presence of complement promotes uptake of L. pneumophila and allows phagocytes to kill about half the ingested organisms

(HORWITZ and SILVERSTEIN 1981a). The surviving organisms are evidently unaffected, as they multiply in monocytes at the same rate as organisms that enter phagocytes in the absence of antibody (HORWITZ and SILVERSTEIN 1981b).

2.2 Receptors and Ligands

L. pneumophila phagocytosis is mediated by a three component receptor-ligand-acceptor molecule system consisting of complement receptors on human mononuclear phagocytes (CR1 and CR3 on monocytes), fragments of complement component C3 fixed to the bacterial surface, and the major outer-membrane protein (MOMP) of *L. pneumophila*, the C3 acceptor molecule in the bacterial outer membrane (PAYNE and HORWITZ 1987, BELLINGER-KAWAHARA and HORWITZ 1990) (Fig. 1). C3 fixes selectively to the MOMP by the alternative pathway of complement activation (BELLINGER-KAWAHARA and HORWITZ 1990). Components of this phagocytic system have been studied in relative isolation. MOMP incorporated into unilamellar liposomes fixes C3 avidly, in contrast to liposomes not containing MOMP, and the liposome-MOMP-C3 complexes are readily ingested by human monocytes and incorporated into membrane-bound phagosomes (BELLINGER-KAWAHARA and HORWITZ 1990).

Complement receptors of mononuclear phagocytes play a general role in mediating phagocytosis of intracellular pathogens (PAYNE and HORWITZ 1987). In addition to *L. pneumophila*, they have been shown to mediate uptake of several intracellular bacteria including *Mycobacterium tuberculosis* (SCHLESINGER et al. 1990), *Mycobacterium leprae* (SCHLESINGER and HORWITZ 1990a, b), *Mycobacterium*

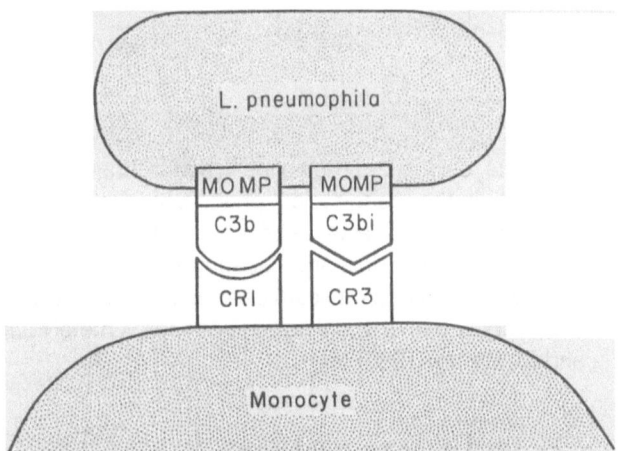

Fig. 1. Three-component phagocytic system mediating uptake of *L. pneumophila*. Phagocytosis by human monocytes is mediated by phagocyte complement receptors (*CR1, CR3*), fragments of complement component C3 (*C3b, C3bi*), and major outer-membrane protein (*MOMP*) in the bacterial outer membrane. (from HORWITZ 1993)

avium (BERMUDEZ et al. 1991), *Listeria monocytogenes* (DREVETS and CAMPBELL 1991), *Bordetella pertussis* (SAUKKONEN et al. 1991), the fungus *Histoplasma capsulatum* (BULLOCK and WRIGHT 1987), and the parasites *Leishmania donovani* (BLACKWELL et al. 1985); WILSON and PEARSON 1987) and *Leishmania major* (MOSSER and EDELSON 1985). It has been suggested that the complement receptor pathway allows safe passage of pathogens into the mononuclear phagocyte (PAYNE and HORWITZ 1987), since ligation of complement receptors by particles coated with fragments of complement component C3 does not trigger an oxidative burst and the release of toxic oxygen metabolites (WRIGHT and SILVERSTEIN 1983; YAMAMOTO and JOHNSTON 1984).

2.3 Membrane Sorting

During phagocytosis of *L.pneumophila*, plasma membrane components of the monocyte are rapidly sorted into or out of the nascent phagosome (CLEMENS and HORWITZ 1992). Complement receptor CR3, which mediates uptake of the organism, is concentrated in the nascent phagosome. In contrast, class-I and -II MHC molecules are excluded from the nascent phagosome. Paralleling the sorting of these receptors, proteins of the plasma membrane can either be concentrated (e.g., 5'-nucleotidase) or excluded from (e.g., alkaline phosphatase) the nascent phagosome. Thus, even before the nascent phagosome is formed, the phagosomal membrane has been remodeled, such that it differs markedly from the plasma membrane from which it is derived.

3 The Phagosome

3.1 Organelle Recruitment

After phagocytosis, *L. pneumophila* resides and multiplies in a membrane-bound phagosome throughout its life cycle in the mononuclear phagocyte. Within a few moments of entry, the phagosome undergoes a series of remarkable interactions sequentially with smooth vesicles, mitochondria, and ribosomes of the host cell (HORWITZ 1983a). After 4–8 h, a ribosome-lined replicative vacuole is formed. The organism multiplies within the vacuole with a doubling time of approximately 2 h at mid-log phase (HORWITZ and SILVERSTEIN 1980) (Fig. 2).

3.2 Inhibition of Phagosome–Lysosome Fusion and Acidification

Two noteworthy features of the *L. pneumophila* phagosome are that it does not fuse with lysosomes (HORWITZ 1983b) and it does not become highly acidified (HORWITZ and MAXFIELD 1984). In these respects, the *L. pneumophila* phagosome

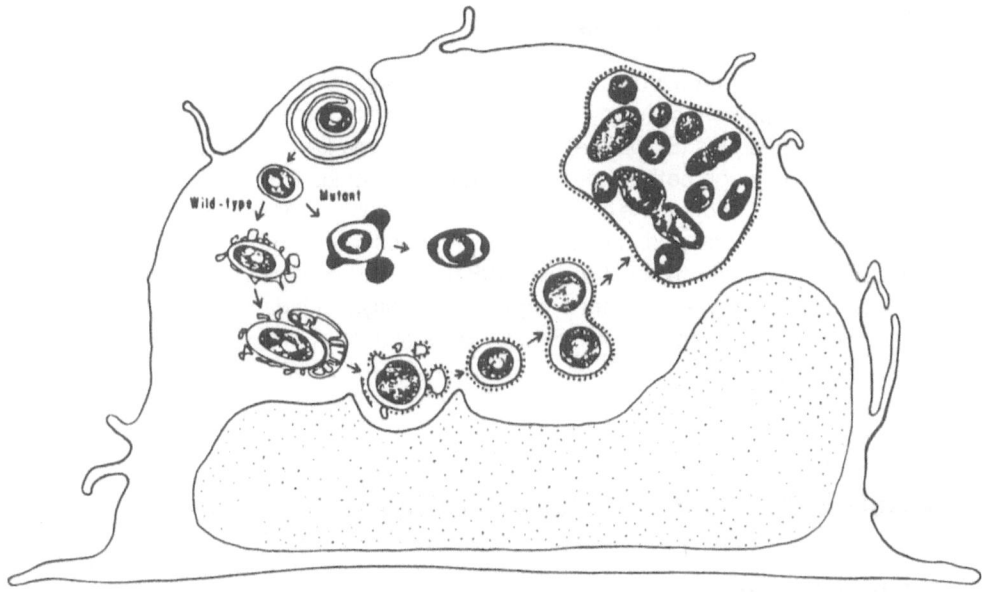

Fig. 2. Intracellular pathways of wild-type and avirulent mutant *L. pneumophila* in mononuclear phagocytes. Both the wild-type and avirulent mutant (25D) enter by coiling phagocytosis, but thereafter their pathways diverge. The wild-type phagosome interacts sequentially with smooth vesicles, mitochondria, and ribosomes, whereas the avirulent mutant does not recruit these organelles. The phagosome of the wild-type *L. pneumophila*, but not the avirulent mutant, avoids fusion with lysosomes. The wild-type organism multiplies in the ribosome-lined phagosome until the monocyte becomes full of bacteria and disintegrates. The mutant survives but does not multiply in its phagolysosome, and it does not destroy the monocyte. (from HORWITZ 1989)

resembles those of *Mycobacterium tuberculosis* (ARMSTRONG and HART 1971; CROWLE et al. 1991) and *Toxoplasma gondii* (JONES and HIRSCH 1972; SIBLEY et al. 1985). However, recent studies have indicated that the *L. pneumophila* and *M. tuberculosis* phagosomes are distinct in other respects (see below).

3.3 Molecular Characteristics

After phagocytosis, the *L. pneumophila* phagosome rapidly becomes devoid of plasma membrane, endosomal class-II vesicles, and lysosomal markers including class-I and -II MHC molecules, transferrin receptors, and lysosome-associated membrane glycoproteins (CD63, LAMP1, and LAMP2), as studied by the cryo-section immunogold technique (CLEMENS and HORWITZ 1993, 1995). This is in contrast to the *M. tuberculosis* phagosome, which exhibits delayed clearance of class-I MHC molecules, frequent staining for class-II MHC molecules and the endosomal marker transferrin receptor, and relatively weak but frequent staining for the late endosomal/lysosomal markers CD63, LAMP1, and LAMP2 (CLEMENS and HORWITZ 1995). Thus, the intraphagosomal pathway, i.e., the pathway followed by pathogens such as *L. pneumophila* and *M. tuberculosis* that inhibit phagosome-lysosome fusion, is heterogeneous.

3.4 Avirulent Mutant Phagosome

The phagosome of an avirulent mutant *L. pneumophila* (Mutant 25D) which is unable to multiply intracellularly in human monocytes (HORWITZ 1987) or cause disease in guinea pigs (BLANDER et al. 1989) differs markedly from the wild-type phagosome, although both enter by coiling phagocytosis (Fig. 2). In contrast to the wild-type phagosome, the avirulent mutant phagosome does not interact with monocyte smooth vesicles, mitochondria, and ribosomes, and it does not inhibit phagosome-lysosome fusion (HORWITZ 1987). Also in contrast to the wild-type phagosome, the avirulent mutant phagosome displays abundant lysosome-associated membrane glycoproteins (D.L. CLEMENS and M.A. HORWITZ unpublished data).

4 Role of Iron in Intracellular Pathogenesis

4.1 Iron Acquisition

L. pneumophila, which has no known siderophores, requires an unusually high concentration of iron for growth on artificial medium. Inside monocytes, *L. pneumophila* evidently readily obtains iron from the intermediate labile iron pool of the host cell (BYRD and HORWITZ 1989, 1991a, 1991b, 1993). This pool derives its iron primarily from iron-transferrin via transferrin receptors, iron-lactoferrin via lactoferrin receptors, and the iron storage protein ferritin, which recycles iron to the pool. Substances that directly or indirectly deplete the intermediate iron pool inhibit *L. pneumophila* intracellular multiplication. Such substances include (a) interferon gamma, which down-regulates transferrin receptor expression and intracellular ferritin concentration, shutting off the two major sources of iron to the pool (BYRD and HORWITZ 1989, 1993); (b) weak bases, including chloroquine and ammonium chloride, which raise the pH of endocytic vesicles and lysosomes and hence inhibit the pH-dependent release of iron from endocytized transferrin and the pH-dependent proteolysis and release of iron from iron-lactoferrin and ferritin (BYRD and HORWITZ 1991a); and (c) iron chelators, including deferoxamine and apolactoferrin, the latter molecule a host iron-binding protein internalized by specific receptors on the plasma membrane of mononuclear phagocytes (BYRD and HORWITZ 1989, 1991b).

4.2 PMN–Monocyte Cooperation in Host Defense

The capacity of apolactoferrin to inhibit *L. pneumophila* intracellular multiplication has suggested a role for polymorphonuclear leukocytes (PMN) in host defense against *L. pneumophila*. PMN are a prominent histologic feature of the

L. pneumophila-infected lung, but these phagocytes exhibit only a modest capacity for killing *L. pneumophila* even in the presence of anti-*L. pneumophila* antibody and complement (HORWITZ and SILVERSTEIN 1981a). However, PMN are rich in apolactoferrin, and they release this iron chelator at sites of inflammation. This raises the possibility that PMN cooperate with mononuclear phagocytes in host defense against *L. pneumophila* by providing infected mononuclear phago-cytes in the lung with apolactoferrin (BYRD and HORWITZ 1991b). This may be particularly important to host defense early in infection, before a protective cell-mediated immune response is generated (HORWITZ and SILVERSTEIN 1981c; HORWITZ 1983c).

4.3 Iron-containing Proteins of *L. pneumophila*

When *L. pneumophila* is grown in the presence of ^{59}Fe, seven major proteins are labeled (MENGAUD and HORWITZ 1993). The major iron-containing protein (MICP) is a monomeric protein of calculated molecular mass 98, 147 Da. MICP is an aconitase highly homologous with *E. coli* aconitase; however, the protein is much more abundant in *L. pneumophila* than in *E. coli*. MICP is also highly homologous with the human iron-responsive element-binding protein. Why *L. pneumophila* requires such high levels of MICP is unknown.

A second major iron-containing protein has superoxide dismutase (SOD) activity (MENGAUD and HORWITZ 1993). The gene encoding the *L. pneumophila* iron SOD has been cloned (STEINMAN 1992) and has been shown to be essential for viability (SADOSKY et al. 1994).

5 Genetic Analysis of Intracellular Multiplication and Macrophage Killing

5.1 Functional Complementation of an Avirulent Mutant

The avirulent mutant (25D) described in Sect. 3.4 differs from wild-type *L. pneumophila* in two key respects. First, it is unable to multiply intracellularly and second, it is unable to kill host cells, even at multiplicities of infection as high as 10^3: 1 Both of these characteristics may reflect the mutant's inability to direct the formation of an LSP. Indeed, the lack of cytotoxicity of 25D may be a direct consequence of its inability to replicate in a non-LSP compartment.

Functional complementation was used to identify the genes and their prod-ucts that may be important for *Legionella* to successfully direct the formation of the LSP. A genomic library of wild-type Philadelphia-1 was prepared in a plasmid vector and introduced en masse into the avirulent mutant. A simple plaque assay was used to identify bacteria which had received a region of *Legionella* DNA that restored their capacity for intracellular multiplication (the Icm phenotype) and macrophage killing (the Mak phenotype). In this plaque assay, macrophages are allowed to spread confluently on plastic dishes and are infected with bacteria in a small volume. The macrophages are then overlayed with agarose to keep bacteria that subsequently multiply intracellularly and lyse macrophages from swimming far from the initial point of infection. Bacteria that are Icm$^+$ Mak$^+$ produce zones of cytotoxicity surrounding the initial point of infection. Mutant 25D produces no detectable plaques even at high multiplicities of infection. When a population of 25D containing the wild-type library was used to infect macrophage monolayers, plaques were obtained at a frequency of approximately 10^{-5}. It was possible to demonstrate that the ability to form plaques was conferred by a specific plasmid clone, 19b. Introduction of clone 19b restored the ability of 25D to replicate intracellularly and to kill human macrophages derived from the HL- 60 cell line (MARRA et al. 1990, 1992).

Animal studies indicated that the 19b clone also restored the ability of the 25D strain to cause lethal pneumonia in guinea pigs infected via the aerosol route. Examination of monocyte-derived macrophages infected with 25D/19b by transmission electron microscopy indicated that the LSP was formed and that it did not fuse with secondary lysosomes. Organelles were recruited by the LSP, but at late times following infection, many fewer ribosomes were observed surrounding LSPs containing 25D/19b than those containing wild-type *Legionella* (MARRA et al. 1992).

Molecular analysis of the 19b clone indicated that approximately 12 kb present on a plasmid subclone (pAM10) was sufficient to complement the Icm and Mak phenotypes of mutant 25D. Furthermore, pAM10 was found to contain the *dotA* gene and four other open reading frames, *icmWXYZ* (BRAND et al. 1994; BERGER et al. 1994). The *dotA* gene was identified in an independent study by Berger and Isberg, who isolated a series of mutants that were defective in intracellular replication (BERGER and ISBERG 1993). Two classes of mutants were complemented by a plasmid which contained *dotA*. With both classes of *dotA* alleles, organelle traffic in infected cells was distinctly different from that in mononuclear phagocytes infected with wild-type *Legionella* (BERGER and ISBERG 1993). Although organelle traffic is certainly defective in macrophages infected by the 25D mutant, as in *dotA* mutants, the precise genetic differences between the wild-type and the 25D mutant are not known. As described above, the 25D mutant was obtained following serial passage of populations on enriched Muller-Hinton agar, which has subsequently been shown to selectively inhibit the growth of virulent *L. pneumophila* (CATRENICH and JOHNSON 1989). An understanding of exactly how the products of the *dotA/icmWXYZ* locus enable *Legionella* to direct formation of the LSP and restore other characteristics of the wild-type phagosome awaits further investigation of the biochemistry and the cell biology of host cell interactions with well-defined mutants.

5.2 Transposon Mutagenesis and the Identification of Additional Icm Genes

The studies aimed at discovering *Legionella* genes required for intracellular replication described in the previous section focused attention on the *dotA/icmWXYZ* locus. To determine directly if the *dotA/icmWXYZ* region is the only locus required for intracellular multiplication and macrophage killing, or if additional genes are also required, SADOSKY et al. screened a large collection of transposon-induced mutants for those that had lost the ability to kill HL-60-derived macrophages (SADOSKY et al. 1993). A transposon derived from Tn*903*, Tn*903*dII *lacZ*, was found to transpose efficiently in *Legionella* (WIATER et al. 1994), and approximately 4600 independent insertions were collected. Among them, 55 were found to have decreased ability to kill HL-60-derived macrophages (Mak⁻). It was possible to measure the cytotoxicity of the different strains in a tissue culture assay using a tetrazolium dye reduction assay for living adherent cells; the apparent TC_{50} for the wild-type was between 10 and 100 colony-forming units (CFU), whereas the TC_{50} for the 25D mutant was $> 10^5$ CFU. Among the 55 Mak⁻ mutants, the relative TC_{50} values were 10^2 to $> 10^5$ greater than that of the wild type.

It was important to establish whether the 55 Mak⁻ mutants represented a collection of insertions in different genes or were all insertions in the *dotA/icmWXYZ* locus. Genomic DNA preparations from all 55 strains were prepared and cleaved with *Eco*RI, a restriction endonuclease which does not cleave within transposon sequences. When these samples were hybridized with labeled transposon sequences as probes, genomic fragments of different molecular weights were found, indicating that the transposon had inserted in multiple sites, albeit a limited number. It was possible to arrange the 55 mutants into 16 groups based on the identity of the *Eco*RI fragment containing the site of transposon insertion. One group containing 11 members corresponded to the *dotA/icmWXYZ* locus. The other groups were clearly distinct from this locus and must represent other genes whose functions are essential for the ability to kill macrophages.

Representatives from each group were assayed for their ability to be taken up by HL-60-derived macrophages, and all were found to be indistinguishable from wild type in this respect. In addition, all were found to be unable to replicate in macrophages over a 48-h period, indicating that the Mak⁻ phenotype was associated with the Icm⁻ phenotype in all of the mutants. Therefore, it may be difficult to isolate mutants which are unable to multiply intracellularly but retain the Mak⁺ phenotype or mutants that retain the Icm⁺ phenotype but have lost the ability to kill macrophages. These results indicate that cytotoxicity is a direct consequence of intracellular multiplication.

What functions do the many different *icm* loci encode? How many different genes are represented by the Icm mutants? All of the mutants grow normally on standard *Legionella* bacteriologic medium (ACES-buffered charcoal yeast extract). This would seem to make it less likely that major catabolic pathways are defective. Whether any of the mutants are auxotrophic has not been definitively addressed, although all grow on a casamino-acid-based medium. The only secondary phenotype that has been detected is decreased sensitivity to Na^+. The growth of wild-type Legionella is inhibited by 100 mM Na^+ or Li^+. The effect is not due to osmolarity or ionic strength. CATRENICH and JOHNSON showed that variants arise frequently which exhibit increased sodium tolerance, and many of these lose the ability to cause disease or grow in macrophages (CATRENICH and JOHNSON 1989). Indeed, the 25D avirulent mutant is likely the result of selection for sodium tolerance imposed by enriched Muller-Hinton medium. With all of the Tn903dII lacZ insertion mutants, an increased LD_{50} for macrophages is correlated with increased sodium tolerance. It has been shown in many cases that the transposon insertion is the cause of both the Icm/Mak and sodium tolerance phenotypes. Unfortunately, these observations raise more questions than they provide answers. Although the molecular basis of Na^+ sensitivity/resistance is unknown, it is tempting to speculate that the organism may use low Na^+ concentration as an indicator of an intracellular environment.

5.3 Cloning of Genes Potentially Involved in Intracellular Multiplication or Macrophage Killing

A classic approach to identifying potential virulence determinants is to examine those components recognized by the host immune response. Convalescent sera from patients who have recuperated from legionnaires' disease contain antibodies to various bacterial components. These sera were used to screen E. coli colonies of Legionella genomic libraries (Engelberg et al. 1984). This analysis yielded a gene that encodes a 24 000-molecular-weight protein present on the Legionella surface which exhibits prolyl-isomerase activity and resembles a class of proteins found in all prokaryotes and eukaryotes that bind the immuno-suppressant FK-506 (CIANCIOTTO et al. 1989). Site-directed mutants of Legionella that were defective in this protein were found to have a defect in macrophage infection, although they retained the ability to enter macrophages and multiply within them. The gene encoding the protein is called mip, for macrophage infectivity potentiator (CIANCIOTTO et al. 1990). The mip gene is apparently widely distributed in Legionella species and related genera. The precise role of the Mip protein in potentiating infectivity and the relation, if any, of prolyl-isomerase activity to intracellular pathogenesis are unclear at the present time.

Another Legionella protein which is an important antigen during infection and has been shown to be important for the development of a protective cellular immune response in guinea pigs is the major secretory protein (MSP) (BLANDER and HORWITZ 1989). MSP, a Zn^{++}-containing metalloprotease with a broad specificity, is abundant in Legionella culture supernates. Disruption of the MSP structural gene (QUINN and TOMPKINS 1989) (mspA or proA) reduces extracellular protease activity to less than 0.1% of wild-type levels on substrates such as casein but has little if any effect on intracellular multiplication, the ability to kill human macrophages, or the ability to produce lethal pneumonia in guinea pigs

(SZETO and SHUMAN 1989; BLANDER et al. 1990; MOFFATT et al. 1994). The role of the protease during legionnaires' disease in human beings is more difficult to assess. It is perhaps curious that a major determinant of immunity would have such a subtle role in pathogenesis. One should bear in mind, however, that mammals are likely not the longest extant ecological niche for this organism, and MSP may confer a substantial advantage to *Legionella* in circumstances which are unknown at the present time.

6 Conclusions

Intracellular bacterial pathogens have evolved a variety of mechanisms to satisfy their nutritional needs. *L. pneumophila*, which primarily parasitizes protozoons in nature, is also able to replicate in mammalian mononuclear phagocytes. To do this, it modifies the phagosome so that it does not acidify nor fuse with secondary lysosomes. It also rapidly sorts host cell plasma membrane markers out of the phagosome and efficiently blocks interactions of the phagosome with the entire endosomal-lysosomal pathway. It is very likely that other differences between the LSP and conventional phagosomes exist. For example, components of the endoplasmic reticulum have been found surrounding the LSP (M. SWANSON and R.R. ISBERG personal communication), in addition to ribosomes. The significance of this observation in terms of intracellular replication is not yet known, but it illustrates the need to identify both host-derived components of the LSP and the *Legionella*-encoded proteins that direct its formation during and immediately following phagocytosis.

Acknowledgements. This work was supported by National Institutes of Health grant A123549 to Dr. Shuman and grants A122421 and A135275 to Dr. Horwitz. Dr. Shuman is the recipient of a Faculty Research Award (FRA-357) from the American Cancer Society.

References

Armstrong JA, Hart PD (1971) Response of cultured macrophages to *Mycobacterium tuberculosis* with observations on fusion of lysosomes with phagosomes. J Exp Med 134: 713–740
Bellinger-Kawahara C, Horwitz MA (1990) Complement component C3 fixes selectively to the major outer membrane protein (MOMP) of *Legionella pneumophila* and mediates phagocytosis of liposome–MOMP complexes by human monocytes. J Exp Med 172: 1201–1210
Berger KH, Isberg RR (1993) Two distinct defects in intracellular growth complemented by a single genetic locus in *Legionella pneumophila*. Mol Microbiol 7: 7–19
Berger KH, Merriam JJ, Isberg RR (1994) Altered intracellular targeting properties associated with mutations in the *Legionella pneumophila dotA* gene. Mol Microbiol 14: 809–822
Bermudez LE, Young LS, Enkel H (1991) Interaction of *Mycobacterium avium* complex with human macrophages: roles of membrane receptors and serum proteins. Infect Immun 59: 1697–1702

Blackwell JM, Ezekowitz RAB, Roberts MB, Channon JY, Sim RB, Gordon S (1985) Macrophage complement and lectin-like receptors bind *Leishmania* in the absence of serum. J Exp Med 162: 324–331

Blander SJ, Horwitz MA (1989) Vaccination with the major secretory protein of *Legionella pneumophila* induces cell-mediated and protective immunity in a guinea pig model of legionnaires' disease. J Exp Med 169: 691–705

Blander SJ, Breiman RF, Horwitz MA (1989) A live avirulent mutant *Legionella pneumophila* vaccine induces protective immunity against lethal aerosol challenge. J Clin Invest 83: 810–815

Blander SJ, Szeto L, Shuman HA, Horwitz MA (1990) An immuno-protective molecule, the major secretory protein of *Legionella pneumophila*, is not a virulence factor in a guinea pig model of legionnaires' disease. J Clin Invest 86: 817–824

Brand BC, Sadosky AB, Shuman HA (1994) Molecular genetic analysis of the *icm* region in *Legionella pneumophila*. Mol Microbiol 14: 797–808

Bullock WE, Wright SD (1987) Role of the adherence-promoting receptors, CR3, LFA-1, and p150, 95 in binding of *Histoplasma capsulatum* by human macrophages. J Exp Med 165: 195–210

Byrd TF, Horwitz MA (1989) Interferon gamma-activated human monocytes down-regulate transferrin receptors and inhibit the intracellular multiplication of *Legionella pneumophila* by limiting the availability of iron. J Clin Invest 83: 1457–1465

Byrd TF, Horwitz MA (1991a) Chloroquine inhibits the intracellular multiplication of *Legionella pneumophila* by limiting the availability of iron. A potential new mechanism for the therapeutic effect of chloroquine against intracellular pathogens. J Clin Invest 88: 351–357

Byrd TF, Horwitz MA (1991b) Lactoferrin inhibits or promotes *Legionella pneumophila* intracellular multiplication in nonactivated and interferon gamma-activated human monocytes depending upon its degree of iron saturation. Iron-lactoferrin and nonphysiologic iron chelates reverse monocyte activation against *Legionella pneumophila*. J Clin Invest 88: 1103–1112

Byrd TF, Horwitz MA (1993) Regulation of transferrin receptor expression and ferritin content in human mononuclear phagocytes: coordinate upregulation by iron-transferrin and down-regulation by interferon gamma. J Clin Invest 91: 969–976

Catrenich CE, Johnson W (1989) Characterization of the selective inhibition of growth of virulent *Legionella pneumophila* by supplemented Mueller-Hinton medium. Infect Immun 57: 1862–1864

Chang KP (1979) *Leishmania donovani* promastigote–macrophage surface interactions in vitro. Exp Parasitol 48: 175–189

Cianciotto NP, Einstein BI, Mody CH, Toews GB, Engelberg NC (1989) A *Legionella pneumophila* gene encoding a species-specific surface protein potentiates initiation of intracellular infection. Infect Immun 57: 1225–1262

Cianciotto NP, Einstein BI, Mody CH, Engelberg NC (1990) A mutation in the mip gene results in attenuation of *Legionella pneumophila* virulence. J Infect Dis 162: 121–126

Clemens DL, Horwitz MA (1992) Membrane sorting during phagocytosis: selective exclusion of MHC molecules but not complement receptor CR3 during conventional and coiling phagocytosis. J Exp Med 175: 1317–1326

Clemens, DL, Horwitz MA (1993) Hypoexpression of major histocompatibility complex molecules on *Legionella pneumophila* phagosomes and phagolysosomes. Infect Immun 61: 2803–2812

Clemens DL, Horwitz MA (1995) Characterization of the *Mycobacterium tuberculosis* phagosome and evidence that phagosomal maturation is inhibited. J Exp Med 181: 257–270

Crowle A, Dahl R, Ross E, May M (1991) Evidence that vesicles containing living virulent *M. tuberculosis* or *M. avium* in cultured human macrophages are not acidic. Infect Immun 59: 1823–1831

Drevets DA, Campbell PA (1991) Roles of complement and complement receptor type 3 in phagocytosis of *Listeria monocytogenes* by inflammatory mouse peritoneal macrophages. Infect Immun 59: 2645–2652

Engelberg NC, Pearlman E, Einstein BI (1984) *Legionella pneumophla* surface antigens cloned and expressed in *E. coli* are translocated to the host cell surface and interact with specific anti-*Legionella* antibodies. J Bacteriol 160: 199–203

Horwitz MA (1983a) Formation of a novel phagosome by the legionnaires' disease bacterium (*Legionella pneumophila*) in human monocytes. J Exp Med 158: 1319–1331

Horwitz MA (1983b) The legionnaires' disease bacterium (*Legionella pneumophila*) inhibits phagosome-lysosome fusion in human monocytes. J Exp Med 158: 2108–2126

Horwitz MA (1983c) Cell-mediated immunity in legionnaires' disease. J Clin Invest 71: 1686–1697

Horwitz MA (1984) Phagocytosis of the legionnaires' disease bacterium (*Legionella pneumophila*) occurs by a novel mechanism: engulfment within a pseudopod coil. Cell 36: 27–33

Horwitz MA (1987) Characterization of avirulent mutant *Legionella pneumophila* that survive but do not multiply within human monocytes. J Exp Med 166: 1310–1328

Horwitz MA (1989) The immunobiology of *Legionella pneumophila*. In: Mounder JW (ed) Intracellular parasitism. CRC Press, Boca Raton, pp 141–156

Horwitz MA (1993) State-of-the-art address.. Toward an understanding of host and bacterial molecules mediating *Legionella pneumophila* pathogenesis. *Legionella*, current status and emerging perspectives. Barbaree J, Breiman R, Dufour AP (eds) In: American Society of Microbiology, Washington DC, pp 55–62

Horwitz MA, Silverstein SC (1981c) Activated human monocytes inhibit the intracellular multiplication of legionnaires' disease bacteria. J Exp Med 154: 1618–1635

Horwitz MA, Maxfield FR (1984) *Legionella pneumophila* inhibits acidification of its phagosome in human monocytes. J Cell Biol 99: 1936–1943

Jones TC, Hirsch JG (1972) The interaction between *Toxoplasma gondii* and mammalian cells. II. The absence of lysosomal fusion with phagocytic vacuoles containing living parasites. J Exp Med 136: 1173–1194

Horwitz MA, Silverstein SC (1980) The legionnaires' disease bacterium (*Legionella pneumophila*) multiplies intracellularly in human monocytes. J Clin Invest 66: 441–450

Horwitz MA, Silverstein SC (1981a) Interaction of the legionnaires' disease bacterium (*Legionella pneumophila*) with human phagocytes. I. *L. pneumophila* resists killing by polymorphonuclear leukocytes, antibody, and complement. J Exp Med 153: 386–397

Horwitz MA, Silverstein SC (1981b) Interaction of the legionnaires' disease bacterium (*Legionella pneumophila*) with human phagocytes. II. Antibody promotes binding of *L. pneumophila* to monocytes but does not inhibit intracellular multiplication. J Exp Med 153: 398–406

Jones TC, Yeh S, Hirsch JG (1972) The interaction between *Toxoplasma gondii* and mammalian cells. I. Mechanism of entry and intracellular fate of the parasite. J Exp Med 136: 1157–1172

Marra A, Horwitz MA, Shuman HA (1990) The HL-60 model for the interaction of human macrophages with the legionnaires' disease bacterium. J Immunol 144: 2738–2744

Marra A, Blander SJ, Horwitz MA, Shuman HA (1992) Identification of a *Legionella pneumophila* locus required for intracellular multiplication in human macrophages. Proc Natl Acad Sci USA 89: 9607–9611

Mengaud, J, MA Horwitz (1993) The major iron-containing protein of *Legionella pneumophila* is an aconitase homologous with the human iron-responsive element-binding protein. J Bacteriol 175: 5666–5676

Moffat JF, Edelstein PH, Regula DP Jr, Cirillo JD, Tompkins LS (1994) Effects of an isogenic Zn-metalloprotease-deficient mutant of *Legionella pneumophila* in a guinea-pig pneumonia model. Mol Microbiol 12: 693–705

Mosser DM, Edelson PG (1985) The mouse macrophage receptor for C3bi (CR3) is a major mechanism in the phagocytosis of *Leishmania* promastigotes. J Immunol 135: 2785–2789

Nogueira N, Cohn ZA (1976) *Trypanosoma cruzi*: mechanism of entry and intracellular fate in mammalian cells. J Exp Med 143: 1402–1420

Payne NR, Horwitz MA (1987) Phagocytosis of *Legionella pneumophila* is mediated by human monocyte complement receptors. J Exp Med 166: 1377–1389

Quinn FD, Tompkins LS (1989) Analysis of a cloned sequence of *Legionella pneumophila* encoding a 38-kDa metalloprotease possessing haemolytic and cytotoxic activities. Mol Microbiol 3: 797–805

Sadosky AB, Wiater LA, Shuman HA (1993) Identification of *Legionella pneumophila* genes required for growth within and killing of human macrophages. Infact Immun 61: 5361–5373

Sadosky AB, Wilson JW, Steinman HM, Shuman HA (1994) The iron superoxide dismutase of *Legionella pneumophila* is essential for viability. J Bacteriol 176: 3790–3799

Saukkonen K, Cabellos C, Burroughs M, Prasad S, Tuomanen E (1991) Integrin-mediated localization of *Bordetella pertussis* within macrophages: role in pulmonary colonization. J Exp Med 173: 1143–1149

Schlesinger LS, Horwitz MA (1990a) Phagocytosis of leprosy bacilli is mediated by complement receptors CR1 and CR3 on human monocytes and complement component C3 in serum. J Clin Invest 85: 1304–1314

Schlesinger LS, Horwitz MA (1990b) Complement receptors and complement component C3 mediate phagocytosis of *Mycobacterium tuberculosis* and *Mycobacterium leprae*. Int J Lepr 58: 200–201

Schlesinger LS, Bellinger-Kawahara CG, Payne NR, Horwitz MA (1990) Phagocytosis of *Mycobacterium tuberculosis* is mediated by human monocyte complement receptors and complement component C3. J Immunol 144: 2771–2780

Sibley LD, Weidner E, Krahenbuhl JL (1985) Phagosome acidification blocked by intracellular *Toxoplasma gondii.* Nature 315: 416–419

Steinman, HM (1992) Construction of an *Escherichia coli* K-12 strain deleted for manganese and iron superoxide dismutase genes and its use in cloning the iron superoxide dismutase gene of *Legionella pneumophila.* Mol Gen Genet 232: 427–430

Stevens DR, Moulton JE (1978) Ultrastructural and immunological aspects of the phagocytosis of *Trypanosoma brucei* by mouse peritoneal macrophages. Infect Immun 19: 972–982

Szczepanski A, Fleit HB (1988) Interaction between *Borrelia burgdorferi* and polymorphonuclear leukocytes. Phagocytosis and the induction of the respiratory burst. Ann NY Acad Sci 539: 425–428

Szeto L, Shuman HA (1990) The major secreted protein (MSP) of *Legionella pneumophila*, a protease, is not required for intracellular growth or host cell killng. Infect Immun 58: 2585–2592

Tanowitz H, Wittner M, Kress Y, Bloom B (1975) Studies of in vitro infection by *Trypanosoma cruzi.* I. Ultrastructural studies on the invasion of macrophages and L-cells. Am J Trop Med Hyg 25: 25–33

Wiater LA, Sadosky AB, Shuman HA (1994) Transposon mutagenesis of *Legionella pneumophila* with *Tn903dlllacZ*: identification of a growth phase-regulated pigmentation gene. Mol Microbiol 11: 641–653

Wilson ME, Pearson RD (1987) Roles of CR3 and mannose receptors in the attachment and ingestion of *Leishmania donovani* by human mononuclear phagocytes. Infect Immun 56: 363–369

Wright SD, Silverstein SC (1983) Receptors for C3b and C3bi promote phagocytosis but not release of toxic oxygen from human phagocytes. J Exp Med 158: 2016–2023

Wyrick PB, Brownridge EA (1978) Growth of *Chlamydia psittaci* in macrophages. Infect Immun 19: 1054–1060

Yamamoto K, Johnston RB jr (1984) Dissociation of phagocytosis from stimulation of the oxidative metabolic burst in macrophages. J Exp Med 159: 405–416

Subject Index

Current Topics in Microbiology and Immunology

Volumes published since 1989 (and still available)

Vol. 187: **Rupprecht, Charles E.; Dietzschold, Bernhard; Koprowski, Hilary (Eds.):** Lyssaviruses. 1994. 50 figs. IX, 352 pp. ISBN 3-540-57194-9

Vol. 188: **Letvin, Norman L.; Desrosiers, Ronald C. (Eds.):** Simian Immunodeficiency Virus. 1994. 37 figs. X, 240 pp. ISBN 3-540-57274-0

Vol. 189: **Oldstone, Michael B. A. (Ed.):** Cytotoxic T-Lymphocytes in Human Viral and Malaria Infections. 1994. 37 figs. IX, 210 pp. ISBN 3-540-57259-7

Vol. 190: **Koprowski, Hilary; Lipkin, W. Ian (Eds.):** Borna Disease. 1995. 33 figs. IX, 134 pp. ISBN 3-540-57388-7

Vol. 191: **ter Meulen, Volker; Billeter, Martin A. (Eds.):** Measles Virus. 1995. 23 figs. IX, 196 pp. ISBN 3-540-57389-5

Vol. 192: **Dangl, Jeffrey L. (Ed.):** Bacterial Pathogenesis of Plants and Animals. 1994. 41 figs. IX, 343 pp. ISBN 3-540-57391-7

Vol. 193: **Chen, Irvin S. Y.; Koprowski, Hilary; Srinivasan, Alagarsamy; Vogt, Peter K. (Eds.):** Transacting Functions of Human Retroviruses. 1995. 49 figs. IX, 240 pp. ISBN 3-540-57901-X

Vol. 194: **Potter, Michael; Melchers, Fritz (Eds.):** Mechanisms in B-cell Neoplasia. 1995. 152 figs. XXV, 458 pp. ISBN 3-540-58447-1

Vol. 195: **Montecucco, Cesare (Ed.):** Clostridial Neurotoxins. 1995. 28 figs. XI., 278 pp. ISBN 3-540-58452-8

Vol. 196: **Koprowski, Hilary; Maeda, Hiroshi (Eds.):** The Role of Nitric Oxide in Physiology and Pathophysiology. 1995. 21 figs. IX, 90 pp. ISBN 3-540-58214-2

Vol. 197: **Meyer, Peter (Ed.):** Gene Silencing in Higher Plants and Related Phenomena in Other Eukaryotes. 1995. 17 figs. IX, 232 pp. ISBN 3-540-58236-3

Vol. 198: **Griffiths, Gillian M.; Tschopp, Jürg (Eds.):** Pathways for Cytolysis. 1995. 45 figs. IX, 224 pp. ISBN 3-540-58725-X

Vol. 199/I: **Doerfler, Walter; Böhm, Petra (Eds.):** The Molecular Repertoire of Adenoviruses I. 1995. 51 figs. XIII, 280 pp. ISBN 3-540-58828-0

Vol. 199/II: **Doerfler, Walter; Böhm, Petra (Eds.):** The Molecular Repertoire of Adenoviruses II. 1995. 36 figs. XIII, 278 pp. ISBN 3-540-58829-9

Vol. 199/III: **Doerfler, Walter; Böhm, Petra (Eds.):** The Molecular Repertoire of Adenoviruses III. 1995. 51 figs. XIII, 310 pp. ISBN 3-540-58987-2

Vol. 200: **Kroemer, Guido; Martinez-A., Carlos (Eds.):** Apoptosis in Immunology. 1995. 14 flgs. XI, 242 pp. ISBN 3-540-58756-X

Vol. 201: **Kosco-Vilbois, Marie H. (Ed.):** An Antigen Depository of the Immune System: Follicular Dendritic Cells. 1995. 39 figs. IX, 209 pp. ISBN 3-540-59013-7

Vol. 202: **Oldstone, Michael B. A.; Vitković, Ljubiša (Eds.):** HIV and Dementia. 1995. 40 figs. XIII, 279 pp. ISBN 3-540-59117-6

Vol. 203: **Sarnow, Peter (Ed.):** Cap-Independent Translation. 1995. 31 figs. XI, 183 pp. ISBN 3-540-59121-4

Vol. 204: **Saedler, Heinz; Gierl, Alfons (Eds.):** Transposable Elements. 1995. 42 figs. IX, 234 pp. ISBN 3-540-59342-X

Vol. 205: **Littman, Dan. R. (Ed.):** The CD4 Molecule. 1995. 29 figs. XIII, 182 pp. ISBN 3-540-59344-6

Vol. 206: **Chisari, Francis V.; Oldstone, Michael B. A. (Eds.):** Transgenic Models of Human Viral and Immunological Disease. 1995. 53 figs. XI, 345 pp. ISBN 3-540-59341-1

Vol. 207: **Prusiner, Stanley B. (Ed.):** Prions Prions Prions. 1995. 42 figs. VII, 163 pp. ISBN 3-540-59343-8

Vol. 208: **Farnham, Peggy (Ed.):** Transcriptional Control of Cell Growth. 1995. 17 figs. X, 141 pp. ISBN 3-540-60113-9